# THE ULTIMATE
# ENCYCLOPEDIA OF
# BEER

© 2014 by Bill Yenne

This edition published in 2019 by Crestline, an imprint of The Quarto Group
142 West 36th Street, 4th Floor
New York, NY 10018 USA
**T** (212) 779-4972 **F** (212) 779-6058
**www.QuartoKnows.com**

First published in 2014 by Race Point Publishing, an imprint of The Quarto Group, 142 West 36th Street, 4th Floor, New York, NY 10018, USA.

Crestline titles are also available at discount for retail, wholesale, promotional, and bulk purchase. For details, contact the Special Sales Manager by email at specialsales@quarto.com or by mail at The Quarto Group, Attn: Special Sales Manager, 100 Cummings Center Suite 265D, Beverly, MA 01915, USA.

10 9 8 7 6 5 4 3 2 1

ISBN: 978-0-7858-3752-7

Photography credits can be found with individual images with details on page four.

Previously published as *Beer: The Ultimate World Tour.*

Printed in China

# THE ULTIMATE ENCYCLOPEDIA OF

# BEER

## A COMPLETE GUIDE TO THE BEST BEERS FROM AROUND THE WORLD

### BILL YENNE

4

## Acknowledgements and Sources

The photos in the book were mostly supplied through the media galleries of the breweries as credited on the pages herein. Additional technical information, including ABV, stylistic data, etc. was acquired through the substantial database provided online by *Beer Advocate*.

In addition, the author would like to thank the following specific breweries and people for supplying photographs and research information: Aass Bryggeri, through the courtesy of Synnøve Wisløff Samuelsen; Alaskan Brewing courtesy of Luke Bauer; Alchemist Brewery courtesy of Jess Graham; Allagash Brewing Company courtesy of Mathew Trogner and Rob Tod; Asahi courtesy of Aiko Shima; August Schell Brewery courtesy of Kyle Marti; Baladin courtesy of Fabio "Moz" Mozzone; Bavaria (Netherlands) courtesy of Inge van der Heijden; Bear Republic courtesy of Kate Gratto-Bachman and Richard Norgrove; Big Sky Brewing photos by Loren Moulton; Birra Menabrea courtesy of Marco Grandi; Blue Mountain courtesy of Jessica Bullard; Boston Beer Company courtesy of Suzanne Pace; Brasserie Lefèbvre courtesy of Cécile Fontaine; Brasserie Thiriez courtesy of Daniel Thiriez; Bräuerei Aying Franz Inselkammer KG courtesy of Julia Jacob; Bräuerei C&A Veltins courtesy of Julia Buchheister; Bräuunion courtesy of Carina Maurer; Bryggeriet Djaevlebryg courtesy of Per Olaf Huusfeldt; Captain Lawrence Brewing courtesy of Aaron Pozit; Castlemaine Perkins courtesy of Karen McKeering; Central City courtesy of Tim Barnes; Cheval Blanc courtesy of Luc Senécal; Chimay courtesy of Manneken-Brussel Imports; Cigar City courtesy of Justin Clark and Geiger Powell; Deschutes Brewery courtesy of Marie Melsheimer; Diageo courtesy of Hazel Chu, Killian Keys and Rhonda Evans; Diamond Knot Craft Brewing courtesy of Sherry Jennings; Dinkelacker-Schwabenbräu courtesy of Sebastian Gerlach; Dock Street Brewery courtesy of Marilyn Candeloro; Dogfish Head courtesy of Tim Parrott; Feldschlösschen Getränke courtesy of Sabine Strehle; Firestone Walker courtesy of Erica Nolan; Florida Ice and Farm courtesy of Alejandra Zuniga; Flying Dog Brewery courtesy of Erin Weston; Fullers courtesy of Zoe Barb; Georgetown Brewing courtesy of Ingrid Bartels; Gritty McDuff's courtesy of Thomas Wilson; Heineken Hungária courtesy of Zita Szederkényi; Heineken Italia courtesy of Giorgia Bazurli; Hill Farmstead Brewery, Bob M. Montgomery Images; Holgate Brewhouse courtesy of Natasha Holgate; Jennings Brewery courtesy of Karen Varty; Karl Strauss Brewing Company courtesy of Melody Daversa and Chase Kauf; Kauai Island Brewing courtesy of Bret Larson; Kiuchi Brewery courtesy of Harina Katsuyama; Kompania Piwowarska courtesy of Pavel Kwiatkowski; Koningshoeven via Lanny Hoff at Artisanal Imports; Krombacher courtesy of Julia Messerschmidt; Kunstmann courtesy of Paula Benitez; Lagunitas Brewing Company courtesy of Ryan Tamborski and Jeremy Marshall; Les Brasseurs du Nord/Bière Boréale courtesy of Isabelle Charest; Lone Tree courtesy of David Shire; Mad River Brewing courtesy of Tera Spohr; Marston's courtesy of Alice Holden-Brown; Meantime courtesy of Lanny Hoff at Artisanal Imports; Merchant du Vin courtesy of Charles Finkel and Craig Hartinger; Mikkeller courtesy of Robert Ureke; Moosehead courtesy of Elizabeth Harroun; Mythos Brewery courtesy of Eleni Dessipri; Nelson Brewing courtesy of Chad Hansen; New Belgium Brewing courtesy of Bryan Simpson; Nils Oscar courtesy of Karin Wikström at TNF Inexa AB; Nøgne Ø courtesy of Tor Jessen; Nørrebro Bryghus courtesy of Tine Pedersen; Odell Brewing Company courtesy of Amanda Johnson-King; Oettinger Bräuerei courtesy of Jana Gebicke; Ottakringer Bräuerei courtesy of Gabriele Grossberger; Paderborner Bräuerei courtesy of Peter Böhling; Pike Brewery courtesy of Charles Finkel; Ringwood courtesy of Rachel Barnett; Rock Art Brewery courtesy of Renée Nadeau; Royal Unibrew courtesy of Louise Kapel; Russian River Brewing courtesy of Natalie Cilurzo and Vinnie Cilurzo; SABMiller courtesy of Monica Ramirez and Rebecca Spargo; SABMiller Africa courtesy of Julie Honiball; SABMiller Poland courtesy of Pavel Kwiatkowski; Sharp's courtesy of Daniel Baird; Sierra Nevada courtesy of Ryan Arnold; Sinebrychoff courtesy of Timo Mikkola; Spoetzl courtesy of Eric Webber at McGarrah Jessee; Steam Whistle courtesy of Sybil Taylor; Stevens Point Brewery courtesy of Julie Birrenkott; St.. Feuillien courtesy of Sabine Friart; Stone Brewing photos by Studio Schulz; Summit Brewing courtesy of Chip Walton; Svijany courtesy of Petra Winklerova; Tennent's courtesy of Vicki Byers; Tröegs Brewery courtesy of Jeff Herb; Trumer Bräuerei courtesy of Lars Larson and Mark Mattson; Tun Tavern courtesy of Diane Tharp; U Fleku courtesy of F. Jordak; US Beverage courtesy of Georgia Homsany and Justin Fisch; Wellington Brewery courtesy of Brad McInerney; West Brewery courtesy of Ruth Oliver; and finally, the Wye Valley Brewery courtesy of Lizzie Davidson.

# Contents

# Preface

## by Fergal Murray
### Master Brewer, Guinness, Dublin

I have worked in the world of beer for nearly 30 years. I was taught by the masters of their craft in a brewery where knowledge and skill were handed down through more than two centuries, but more than anything, I gained from them a sense of passion for what it takes to be a brewer.

During my career, I have brewed beer on two continents, and I have traveled to over 50 different nations on every continent. I would guess that I have had at least one beer in each of these countries. I suppose I've probably been in more bars across more nations than most of my fellow beer lovers. I've been in brewhouses and bottling halls all over the world and I've witnessed a lot of passion and dedication to the brewing of great beer.

For all brewers, the four basic ingredients—water, malt, hops, and yeast—are fundamentally the only way to make great beer. As I have learned, and as I advise everyone who brews, a good brewer starts with a clear and

precise process, practices a recipe, and continues to practice the same recipe to be consistent and to maintain and improve the quality. As I often say, watch your hygiene, mind your yeast, and cherish the goodness that comes from the malt base. As an old brewer once told me, "get it right in the brew house, and with good yeast, everything else is just time."

However, for a brewer to become a master brewer is to exude a passion for our craft, and to put heart and soul into our beer.

I've have had some extraordinary beer experiences, made remarkable by great locations, great breweries, and unique events and situations to celebrate. I've savored my beer in exclusive high-end bars, in street corner pubs, in beach taverns, and in jungle shacks.

I fondly remember one atrociously hot sweltering night in Lagos many years ago, just after another military coup. We got into one of those traffic jams you can only get into in Africa, where cars, buses, trucks, cows, goats, sheep,

people, and all of God's creations merge, and chaos prevails. (When this happens, you do not try to think how to solve the problem—it can't be done.) This night, two colleagues and I, with our drivers, had gotten out the brewery gate and had entered the fray. I remember that when we realized we had not moved, and the air conditioning in the cars was beginning to falter, we got out of our cars to survey the scene. My colleagues and I, seasoned in Africa, realized that our situation was hopeless.

At that moment, we saw a table beneath a canvas-roofed establishment marked with a poster of a bottle of beer. Most important, we saw that they had stock in a cold bucket. We entered to the surprise of the young woman, who up until then had probably never experienced three expats attempting to sit on her makeshift bench beside the street, and asking for a beer. From that moment on, one of the great beer experiences of my life unfolded. We were an attraction. Soon the corner bar was

overwhelmed with onlookers. It turned into one of those occasions when the one common desire for cold beer transcended all thoughts and concerns. The stories, the fun, the camaraderie were all amazing. Eventually the traffic subsided, and everyone went on his way, but after that, on our way home, we regularly went back to that same table and that cold bucket of beer.

Did you know that next to water and tea, beer is the beverage most consumed by human beings across the globe? More that 200 billion liters are consumed each year.

Never has beer, and the enjoyment of beer, been more diverse, more active than it is today. Even as traditional beer markets are declining, new, emerging markets are growing, along with innovative brands and fabulous craft brewers developing everywhere.

I read a document recently from one of the brewers' associations, which classified over 140 different styles of beer. I know this is growing, and with so much diver-

sity, and so many amazing fantastic tastes, the world of beer has never been more exciting.

As an adorer of great beers, I have found that experiencing great beers is more than just the beer itself. We must have a great bar, a place where it just feels right. Whether it may be a street bar, a high-end, five-star, state-of-art bar, or an iconic, history-laden pub, it just has to have something magic about it. Then we need a great bartender, a bartender who knows how to craft

**Above: Fergal Murray holds a pint poured at the tap in the St. James's Gate kegging facility in Dublin. This is as close to the source as it gets.** (Bill Yenne photo)

and serve a beer so that looks absolutely gorgeous in every way—because it is with our eyes that first savor our beer.

Only then do we take pleasure in our reward, when all the wonderful flavors and textures awaken our senses and bring to life the beer, a refreshing experience that enhances life's journey.

# Introduction
## by Bill Yenne

This book is both a global survey and a global celebration of beer. In this book, we speak of an archetypical beer traveler, who may be on the ground in a local or foreign land, or in an armchair with a selection of beers imported from around the globe. Our book is organized regionally, because regions have their own characteristic brewing culture and heritage of beer styles. It is designed for people who enjoy beer as well as the cultural nuances of brewing history.

Our beer traveler's journey circumnavigates the globe, beginning in Ireland, then moves eastward across northern Europe to Russia. From here, we turn south and travel westward into the heart of Europe. Next, we turn south through southern Europe into the Middle East and Africa, before traveling eastward to Asia. Then, we move south into Australia and New Zealand before crossing the South Pacific to Latin America. Traveling to North America, we visit Canada, then wind up our journey by moving westward

Above: The author enjoying a pint at a favorite pub. (Paul Waters photo)

across the United States. The beer traveler's last stop is the westernmost brewery in the world.

Beer is an ancient beverage, brewed by the Sumerians, enjoyed by the pharaohs, and produced in countless civilizations from the arctic to the equator and beyond since before there was a known written record of human progress.

Professor Solomon Katz at the University of Pennsylvania found Sumerian recipes for beer that date back four millennia, and beer is mentioned

often in ancient Egyptian literature. As H.F. Lutz points out in *Viticulture and Brewing in the Ancient Orient*, Middle Kingdom texts from Beni Hasan "enumerate quite a number of different beers." Among these were a "garnished beer" and a "dark beer." In the *Newsletter of the American Society of Brewing Chemists*, David Ryder notes that Egyptian beer "was also used as a medicine, a tonic for building strength . . . a universal cure for coughs and colds, shortness of breath, problems of the stomach and lungs, and a guard against indigestion!"

It has been suggested that the brewing of beer is at least as old as the baking of bread, and certainly both have been practiced since the dawn of recorded history. Indeed, the art and science of the brewer and those of the baker are quite similar, both involving grain, water and yeast. In fact, the Sumerians baked barley loaves called bappir that could be stored in the dry climate and either eaten as bread or mixed with malted barley to form a mash for brewing.

Beer is mentioned by Xenophon and Aristotle (as quoted by Athenaeus). Among others, the Roman consul and scholar Pliny the Younger estimated that nearly 200 types of beer were being brewed in Europe by the first century. The Latin texts refer to the barley beverages as cerevisia or cerevisium, root words that are still with us in the Spanish and Portuguese words for beer—cerveza and cerveja—as well as in the Latin name for brewer's yeast, *Saccharomyces cerevisiae.*

This is not a book about ancient beer, but it is important to know the origins of our subject.

Each chapter contains a capsule history, because it is useful to know how the beer and brewing culture, as well as the major brands and breweries, evolved over time. This is also an homage to people, of whom there are many, who enjoy collecting labels and brewing ephemera, or breweriana. For them, it is interesting to observe the twists and turns

that specific brands have taken over time.

Hundreds of specific breweries and brands are profiled in detail, with the full knowledge that we have omitted many more, and that such a discussion is a mere snapshot in time and clearly subject to change. In this regard, the book you now hold should be regarded by the beer traveler as a point of departure for a journey. With this book as a portal, there are numerous ways to keep up to date on specific beers and brands. Magazines and websites such as *RateBeer* and *Beer Advocate* contain a depth of constantly updated information.

This is an exciting time in the history of brewing. The large are growing larger through mergers and acquisitions, and the world is growing smaller. Once the world was filled with great national brands, but now the biggest are *international.* Heineken is visible virtually everywhere on earth and controls nearly 200 brands other than its own from Europe to Africa to the Far East, and has a tenth of the global market. In

2002, meanwhile, South African Breweries merged with Miller Brewing, and another global consortium came into being with another tenth of the market. In 2008, InBev, a Brazilian-Belgian conglomerate formed just four years earlier, acquired America's Anheuser-Busch, then the world's largest brewing company, and the resulting AB InBev now owns more than 200 brewing companies, countless brands, and a 20 percent global market share. There is more to come in way of consolidations.

At the opposite end of the spectrum from the venerable giants, the brewing world is crackling with innovation. The interest in imaginative craft brewing, once confined to corners of certain countries, has gone global and is now growing in double digits annually in key markets. In this book, the reader and beer traveler will see how and why. This book is a pictorial presentation of the amazing world of beer in the twenty-first century, and a look at how it evolved over the past centuries, with an indication of where it is going in the future.

### Notes on Measurements and Abbreviations

With the exception of the United States, the brewing industry in most of the world uses hectoliters as a standard measurement. In this book, production and capacity volume is listed in hectoliters for all nations except the United States. The numbers are rounded off throughout, subject to change, and included for historical reference only.

Three very common industry abbreviations are widely used herein without explanation. These are ABV, meaning alcohol by volume; IPA for India Pale Ale; and IBU for International Bitterness Unit, a 100-point scale for measuring the bitterness of beer. On the IBU scale, pale lagers are at around 5, most ales in the 30 to 50 range, IPAs somewhat higher, and Double IPAs closer to 100, the taste-bud threshold.

# What is Beer?

Below: The basic ingredients of beer are malted barley, hops and water, to which yeast is added to complete the transition of the raw materials into beer. (Budvar)

Opposite top: The brewhouse at Alaskan Brewing in Juneau. (Alaskan)

Opposite bottom left: Boiling the wort inside the brew kettle at the Pike Brewery in Seattle. (Pike)

Opposite bottom right: Yeast at work, transforming starch into carbon dioxide and alcohol, inside an open fermenter at Schneider Weisse in Bavaria. (Schneider)

Beer is a beverage originating with grain, in which the flavor of the grain is balanced through the addition of other flavorings. Since the Middle Ages, those flavorings have principally been hops, the intensely-flavored flower of the *humulus lupulus* plant. Originally a preservative as much as a flavoring, hops have been used by brewers for centuries.

Today, a brewer typically starts the process with cereal grains—usually, but not exclusively, barley. The grains are malted, meaning that they are germinated, quickly dried, and often roasted. The extent to which malted grain is roasted imparts color to the beer. The malt is then mashed, or soaked long enough for enzymes to convert starch into fermentable sugar. The mashing takes place in a vessel which is called a mash tun.

Water is then added to the mash to dissolve the sugars, resulting in a thick, sweet liquid called wort. The wort is then boiled in what brewers call a brew kettle. At this point, most brewers add hops. Throughout history, brewers have occasionally added seasonings other than hops to their beer. The Egyptians added flavorings such as fruit and honey, and certain modern beers contain fruit and spices.

Finally, the yeast is added to the cooled, hopped wort, and the mixture is set aside to ferment into beer. During the fermentation process the yeast converts the sugars into alcohol and carbon dioxide. The vast majority of beers in the world today are fermented to between 4- and 8-percent alcohol by volume (ABV), though there are exceptions, and many popular beers in North America and Western Europe today fall in the 9 to 12 percent range.

There are two principal types of yeast used in brewing, though they are not the only ones. Most are fermented with *Saccharomyces pastorianus*, formerly known as *Saccharomyces carlsbergensis*, generically called lager yeast. It is a "bottom-fermenting" yeast because

it does the work of converting starch and sugar into alcohol and carbon dioxide at the bottom of the fermenting vessel. The other major type is top-fermenting *Saccharomyces cerevisiae*, known as ale yeast. The use of spontane-ously-fermenting wild yeast of the genus *Brettanomyces*, once rare outside Belgium's Senne Valley, is now more widespread.

For further detailed information about specific beer styles, the traveler is invited to continue this journey on page 274.

# Ireland

As the traveler passes through the emerald landscape of the Emerald Isle, and perhaps through the doors of any random village pub, there is no question that one particular beer style has an identity that is interwoven with that of Ireland itself. Though football fans may stereotypically favor lager, and urban cocktail lounges cater to the "spirited" set, a pint of Irish stout is Ireland's signature adult beverage. Ireland is not without its signature brand of stout.

Guinness is a name that conjures up images, not only of a black beverage with a creamy head, but of a family, a brand, and a brewery that have been an integral element of Irish commercial and cultural history for more that two centuries. Guinness also derives greatness from its unique sense of place. No other brand of such wide appeal is more endowed with roots that extend so deeply into its native soil. The Irish write folk songs about Guinness. James Joyce mentions it dozens of times.

Arthur Guinness joined the ranks of Irish commercial brewers in 1755, having leased a small brewery in Leixlip, County Kildare. In 1759, Arthur famously made the move to Dublin, where sizable commercial brewing operations dated back to the twelfth century. He leased a

**Below: The entrance to the Guinness Brewery. The original St. James's Gate was perpendicular to this gate. (Bill Yenne photo)**

brewery near St. James's Gate on the same site where the Guinness is brewed to this day. When Arthur died in 1803, his sons inherited the business, with Arthur, known as "the Second Arthur," being the inheriting son, who passed it to his sons. Though the brewery remained in Dublin, the first Arthur's great-grandson, Edward Cecil Guinness, the 1st Earl of Iveagh, took the company public on the London Stock Exchange in 1886 and moved the headquarters to London. He became the richest man in Ireland and the brewery became the largest in the world. In 1997, Guinness merged with Grand Metropolitan Brands to form an entity called Diageo PLC, but Guinness maintains its own brand identity.

Between 1936 and 2005, Guinness operated a second brewery at Park Royal in suburban London. Originally built to serve the United Kingdom mar-

ket, Park Royal eventually eclipsed St. James's Gate in volume. Additional previously independent Irish breweries at places including Kilkenny and Dundalk were acquired and operated well into the twenty-first century. In 1963, Guinness opened a brewery in Nigeria, followed by other joint ventures in Africa and Asia. To this day, Nigeria

Above: Arthur Guinness himself. (Diageo)
Below: Eighth generation grandson Patrick Guinness at the site of Arthur's first brewery in Leixlip. (Bill Yenne photo)

### Three Variations on Guinness Stout

Left: Guinness Extra Stout is the closest to the Guinness Porter originally brewed by Arthur Guinness that is now available. Guinness Extra Stout has been almost completely replaced by Guinness Draught and now represents less than five percent of all Guinness sold worldwide.

Center: First brewed in Dublin in 1802, Guinness Foreign Extra Stout is the original export stout. The fastest growing of all variants, it is the key Guinness product for Caribbean, Africa and Asia. Nigeria is the largest market for Foreign Extra Stout and the third largest market for Guinness worldwide.

Right: The most familiar Guinness variant across the world, Guinness Draught was introduced in 1959. Thanks to the innovation of the widget, launched in 1986, Guinness Draught is now available in cans and bottles, as well as in pubs. The widget helps to form the characteristic smooth and creamy head and create an authentic Guinness every time. The widget works by sending a jet of nitrogen through the body of the beer. This knocks nitrogen and carbon dioxide out of suspension, creating the surge, which results in the creamy head. Guinness Draught in cans or bottles is now available in over 70 countries around the world. (Diageo photos)

to his own notes, Arthur Guinness brewed his last batch of ale at St. James's Gate in April 1799. In the next century, "stout porter" was added, and it became known as "stout."

Today, Guinness brews three variations of stout. These are Guinness Extra Stout, Guinness Foreign Extra Stout, and Draught Guinness. They are generally similar, although Foreign Extra Stout is rated at around 7.5 percent ABV and the others are in the 4 to 5 percent range. Draught Guinness differs from the others as it is delivered on draft with a mixture of nitrogen and carbon dioxide, rather than only carbon dioxide like most other beers. The proportion ranges from 70 to 80 percent nitrogen.

The idea of "nitrogenation" goes back to the 1950s, when the company was looking for an engineering solution to achieving a consistently "perfect" pour. Guinness engineer Michael Ash discovered that the key to creating a stable, creamy head on the beer was a mixture of nitrogen and carbon dioxide within the beer itself. Ash's nitrogenation process was initially rolled out in 1959 for the company's bicentenary and gradually perfected over the next five years.

In 1985, the company introduced a system, developed by Alan Forage and William Byrne, for achieving the Draft Guinness

is the third-largest market for Guinness stout after Ireland and the United Kingdom. The United States is fourth.

Initially, Arthur had only brewed ale at St. James's Gate,

but as English porter became quite popular in Ireland later in the century, he joined the ranks of many of the Dublin and Cork brewers who added porter to their product line. According

effect in a can by inserting a small device called a "widget." Eventually, a similar device was developed for bottled Guinness.

In 1960, correctly foreseeing the rise in importance and popularity of lager beer in traditional Guinness markets, the company decided to stay ahead of the curve by creating its own lager. Guinness hired a German master brewer, Dr. Herman Münder of Cologne, to come to Ireland to create a new lager, and turned the former Great Northern Brewery at Dundalk into a state-of-the-art lager plant. Harp, at 5 percent ABV, has remained an important part of the Guinness portfolio ever since.

Though Guinness has been the dominant brewer in Ireland for as long as anyone now alive can remember, in the nineteenth century it was but one of many. At the time that the two Arthurs were defining their company as a brewer of porter and stout, Cork still rivaled Dublin as Ireland's brewing capital. In Cork, the major brewers at the time were James J. Murphy & Company and the Cork Porter Brewery.

The former brewery was the largest in Ireland at the beginning of the nineteenth century, and a century later, Murphy's was second only to Guinness. These three entities have constituted the "Big Three" of Irish stout for over a century, although the Cork

breweries have not been independent for many years.

As St. James's Gate was located on a site where brewing dated back more than a century, Murphy & Company was founded in 1856 at Lady's Well by the four Murphy brothers, James, William, Jerome and Francis. Located on the hill opposite the brewery, the celebrated Lady's Well was dedicated to the Virgin Mary and believed to possess miraculous properties. The company remained independent until 1983, when it was purchased by Heineken International.

Founded in January 1792 as a partnership between William Beamish, William Crawford, Richard Barrett, and Digby O'Brien, the Cork Porter Brewery became Beamish & Crawford in 1901. They took over the neighboring Southgate Brewery of Lane & Company, which had been founded in 1758. In 1906, St. Stephen's Brewery at Dungarvan was purchased and the Bandon Brewing Company of Allman, Dowden & Company became part of the company in 1914. In turn, Beamish & Crawford itself was purchased by Canadian Breweries Limited (Later Carling O'Keefe Limited) in 1962.

Above: Pints rest between pours on the bar at the Brazen Head, Dublin's oldest pub. (Bill Yenne photo)

**Above: A pint of Murphy's Stout.**
(US Brands)

**Below: A coaster from the Carlow Brewing Company.** (Author's collection)

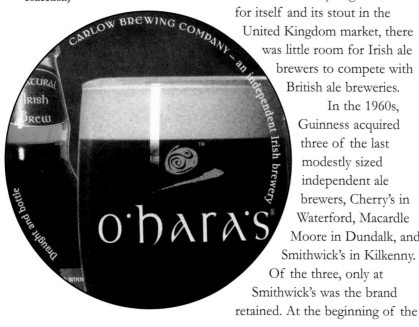

In 1987, Australia-based Elders IXL (later Foster's), then the fourth-largest brewing company in the world, took over Carling O'Keefe and hence the Cork-based brewery as well. This takeover led to the introduction to the Irish market of Foster's Lager. At the same time, Elders also acquired Courage in England. In 1995, Elders sold the brewery to Scottish & Newcastle. When the latter was acquired by Heineken in 2008, both of the big Cork breweries came under the same umbrella.

While the stout breweries and brands survived and flourished, it was a more challenging road for the Irish ale brewers. They had never grown so large and dominant in the domestic or export market. While Guinness established a very large niche for itself and its stout in the United Kingdom market, there was little room for Irish ale brewers to compete with British ale breweries.

In the 1960s, Guinness acquired three of the last modestly sized independent ale brewers, Cherry's in Waterford, Macardle Moore in Dundalk, and Smithwick's in Kilkenny. Of the three, only at Smithwick's was the brand retained. At the beginning of the

twenty-first century, Smithwick's was the oldest surviving brewery in Ireland, having been founded as the St. Francis Abbey Brewery in 1710 by John Smithwick on the site of a fourteenth-century monastic brewery. The Smithwick family battled ups and downs and various hardships and continued brewing at the brewery site for 255 years. Though the Kilkenny brewery is slated by Diageo for closure, the brand remains as one of Diageo's most important.

George Henry Lett's Mill Park Brewery brewed its Enniscorthy Ruby Ale from 1864 to 1956, but a quarter century after the last batch was brewed in Enniscorthy, the American megabrewer Coors bought the rights to the brand and introduced it in the American market as "Killian's Irish Red," naming it after the George Killian Lett, the great-grandson of the founder. It is still brewed, albeit as a lager and not as an ale, by Coors in the United States and by Heineken-owned Pelforth in France.

As in the United States, Ireland experienced the virtual extinction of its tradition of small hometown breweries during the second half of the twentieth century, only to see a resurgence of a new generation of craft breweries at the turn of the twenty-first century. The Carlow Brewing Company, also known as O'Hara's Brewery, was founded in the town of Carlow in 1996 by the O'Hara family. In addition to Irish Stout, the core range includes O'Hara's Irish Red and Curim Gold Wheat Beer/O'Hara's Irish Wheat.

Since 2006, operating in the former Emerald Brewery near Roscommon, Aidan Murphy and Ronan Brennan brew an American-style 4.4 percent ABV pale ale called Galway Hooker. In Allenwood in County Kildare, Trouble Brewing was founded in 2009 by Paul O'Connor, Stephen Clinch and head brewer Thom Prior, who studied brewing and distilling at Heriot-Watt University in Edinburgh. The Trouble portfolio includes Or Golden Ale, Dark Arts Porter, and Sabotage IPA.

For the beer traveler passing through the Irish capital, no visit is complete without a visit to "Dublin's second largest brewery" (after Guinness, of course) and Ireland's first modern brewpub, the Porterhouse. Liam La Hart and Oliver Hughes, whose background had been as pub owners "specialized in importing various beers from around the world with a keen eye on Belgium," opened their brewpub in 1996 in the Temple Bar area in the Dublin city center. With their concept having proven successful, they subsequently opened a second Porterhouse in Dublin, and one

in London's Covent Garden. In 2011, they opened another in Fraunces Tavern, the historic New York establishment where George Washington often dined and enjoyed the occasional glass of porter with his friends and colleagues.

**Above: The Galway Hooker van in Galway Harbor. (Bill Yenne photo)**

**Below: The tap handles at the bar in the Porterhouse in Dublin. (Bill Yenne photo)**

# The United Kingdom

The United Kingdom has long ranked with Germany as one of the two most important brewing nations in Europe, although by style they could not be more different. As lager is the definitive beer style in Germany and most of the world, ale is still the characteristic beer in the United Kingdom. Lager began to outsell ale late in the twentieth century, from 20 percent of the market to 75 percent by the end of the century, but this trend has since reversed. There has always been a larger proportion of ale sales in the United Kingdom than anywhere else on earth.

Thanks to the Campaign for Real Ale (CAMRA), dating back to 1971, consumers not only became more aware of traditional brewers and brewing, they came to take increasing pride in them. "Real ale" in CAMRA's definition is cask-conditioned beer in which fermentation continues to take place in the wooden barrel from which the beer is served. This contrasts with most draft (in Britain, draught) beers, which are filtered and pasteurized before they are put into barrels or kegs.

The legendary British beer commentator Michael Jackson credited CAMRA with saving cask-conditioned ale as a beer style. Producing and enjoying traditional ale pumped directly from unpressurized barrels might well have disappeared, but instead it enjoyed a renaissance, and the practice is

**Below: Enjoying a pint of cask-conditioned Abbot at the local pub. (Greene King)**

**Above: The brewhouse at the West Brewery in the Templeton Building on the Glasgow Green. The brewery produces German-inspired lagers.** (West)

more widespread than ever. At the same time, many smaller traditional brewers were saved, new brewers using traditional methods were born, and consumers have more and better choices.

Interest in cask ale has been increasing in the twenty-first century. Marston's, the largest producer of cask ale in the United Kingdom and the world, reported in 2013 that cask ale accounted for 15 percent of all on-trade (draught) beer poured in the United Kingdom, and 52 percent of all draught ale.

The ranks of the United Kingdom breweries most highly regarded by serious beer enthusiasts are constantly changing, but certain ones seem to percolate to the top of many polls. Though they are profiled later in this chapter, it is worthwhile to make mention of them at the top. Among the newer generation are Harviestoun, Kernel, and Old Chimneys. Among those which have been around for a century or more, and to whom we must tip a hat for their survival in the face of adversity, are J.W. Lees, known for their 11.5 percent ABV Harvest Ale (a barleywine); Samuel Smith's; and two—or three—more that are among the largest of the surviving independent breweries. These are Fuller, Smith & Turner and Wells & Young's, the latter the result of the merger of two of the great old names in British brewing.

United Kingdom brewers have created most of the styles of ale that are known the world over. In the eighteenth century, the great brewing companies began congregating in the city of Burton-upon-Trent. With its abundance of rich hard water and the opening of the railway to Burton in 1839, these brewers had access to the entire United Kingdom market. Burton

Above: Savoring a pint in Cornwall. (Sharps)

Below: An old cask in the cellar at the Westgate Brewery. (Greene King)

became the brewing center of England, and the infinitely imitated English pale ale was born. They created India Pale Ale (IPA), born as a pale ale variant high in alcohol and high in bittering hops, which were essentially there to preserve the beer when exported on long sea voyages to distant parts of the British Empire, such as India. At the turn of the twenty-first century, IPA became the popular mainstay of craft brewers in North America. In the iconic English pub, there were the mild and bitter ale, as well as the more assertive Extra Special Bitter (ESB), all hand pumped from kegs in a manner which would have died out if not for CAMRA, and which is alive and well in the United Kingdom and now a standard practice at many brewpubs in North America.

London can lay claim to being the birthplace of an important

beer style. In 1722, a strong black beer was introduced which used coarse barley and hops that were suited for London's soft water. It is said to have been invented by Ralph Harwood who brewed in Shoreditch, a place within the London borough of Hackney. It mainly appealed to market porters, originally those at Covent Garden, who liked both its body and its price. Hence, it became known as "porter." The real advantage of porter became clear later when it was made in bulk: It did not deteriorate if matured in wooden casks over a long period of time.

From porter, there evolved the blacker, richer *stout* porter, which came to be called stout. As with ale, there are many varieties of stout. On the sweeter side, with more unfermented sugars, sometimes added deliberately, there are sweet stouts and milk stouts. The addition of oats to the mash while brewing produces the especially smooth oatmeal stout. In the late eighteenth century, when British brewers were doing a brisk export business with the Baltic seaports, including St. Petersburg, they created a high alcohol porter called Baltic Porter, which is still brewed locally by many brewers in the region today. Meanwhile, Thrale's brewery in London got a contract with the court of Russia's Catherine the Great herself and created a style known as Russian imperial stout. Long on dark roasted malt character, and high in alcohol, up to the vicinity of 12 percent ABV, imperial stout is still in the portfolio of many British and American craft brewers. Also in the double digits of the ABV scale is barleywine, which is not a wine, of course, but an ale with strength and complexity that ages well and is often cellared.

In the twentieth century, brewing in Britain enjoyed a steady growth until the 1980s. In 1960, total output was 40 million hectoliters, compared with 25 million hectoliters in 1930. The total output reached 65 million hectoliters in 1980, but declined through the end of the century, hitting reaching 36 million in 1992, before climbing to a plateau of around 45 million in the twenty-first century.

Thanks to the change of mood brought about by CAMRA and a microbrewery and brewpub revival, the number of breweries tells an alternate story. In 1960, there were around 300 breweries, compared with 610 in 1930. The number declined to 232 in 1985, but the trend slowly reversed. Since the turn of the century, the resurgence of interest in real ale and craft brewing brought a dramatic increase in the number of breweries. In 2012, the publication of the 40th anniversary edition of CAMRA's *Good Beer Guide* reported 1,009 breweries in operation, the "highest number of breweries in UK for over 70 years [with] 158

Above: Tap handles for pulling cask-conditioned ale. (Greene King)

## British Beer Festivals

The Campaign for Real Ale (CAMRA) organizes numerous festivals throughout the United Kingdom, of which the largest is the **Great British Beer Festival** (GBBF). Founded in 1977, it is held annually in London for five days during the first full week in August. It now attracts more than 500 British and foreign breweries and around 70,000 visitors.

Ten years later, CAMRA inaugurated the annual **National Winter Ales Festival**, usually held in January as a sister festival to the GBBF. It has taken place at various locations, including Manchester, Glasgow, and Burton-upon-Trent.

CAMRA's **Cambridge Beer Festival** is Britain's third largest beer festival, and the oldest, having been held at various locations since 1974. It takes place annually for a week in May, with its current location being on Jesus Green in Cambridge. The **Farnham Beer Exhibition** in Surrey, first held in 1977, is the longest-running annual British beer festival held annually in the same location, predating the GBBF by a few months.

new breweries open for business in just 12 months!"

As in any country, a sizable proportion of British beer sales takes place in taverns and pubs. In Britain, unlike elsewhere, however, most pubs were traditionally "tied houses," meaning they were "tied" to a specific brewery and its beers. In many cases, this involved large brewing companies owning not only hundreds of local pubs, but chains of hotels and restaurants (often referred to as "destination pubs") as well. This makes for a good supply of fresh beer, but it can also be perceived by some as being monopolistic. In 1989, after an investigation by the UK Monopolies and Mergers Commission, the "Beer Orders" were enacted, requiring the presence of "guest ales" at tied houses and restricting the number of tied hoses that a brewery could own.

By the time the Beer Orders were phased out in 2003, the whole landscape of British brewing and pub operations had changed, although most large brewing companies also own, or have owned,

an ancillary hospitality business, operating some mixture of pubs, restaurants, and hotels. In Britain, when one discusses the acquisition of a brewery by another brewery, it is important to remember that it is not merely the acquisition of brewing capacity and/or brands, it is the acquiring of pubs and of handles in other pubs which the acquired brewery does not own outright.

The latter decades of the twentieth century were a period of brewing industry consolidation everywhere in the world, and they gave the United Kingdom its "Big Six," megabrewers who came to dominate the industry from the tied houses to the shelves in the shops. Actually, it was "Big Six Plus One," with the "One" being Guinness, whose brewery at Park Royal near London operated from 1936 to 2005 and was the largest single-site brewery in Britain for many of those years. The list of the Big Six was headed by Bass, followed by Allied, which was the culmination of the mergers of Ind Coope, Allsopp's, Tetley Walker, and Ansells during the early 1960s.

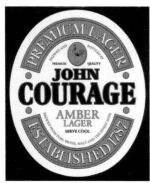

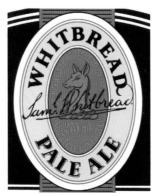

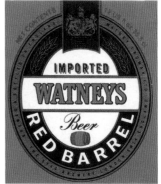

Next, there was Edinburgh-based Scottish & Newcastle, the result of the 1960 merger of Scottish Brewers and Newcastle Breweries. In the next tier were Whitbread, Watney, and Courage.

However, times change, and the phrase "the bigger they come, the harder they fall" comes to mind. Though most of the brand names can still be found if one looks hard enough, none of these companies survived the first decade of the twenty-first century as an independent company. Only Courage remains British-owned, but only as a brand name.

In 1777, William Bass, who ran a cartage business hauling ale to London, opened his own brewery in Burton. William died ten years later, but his son Michael carried on the business, and by 1800 he had a thriving trade exporting beer to the Baltic and Russia as well as to Danzig and German states. Bass was on its way to becoming Britain's largest brewing company. Its red triangle logo is the oldest registered trademark in Britain.

As the British Empire grew, so too did the Bass empire. In 1832, Bass developed its India Pale Ale, which traveled favorably to distant India as well as to Egypt and South Africa. By the Bass centennial in 1877, output had reached a 1.2 million hectoliters annually, and the bottles with the red triangle were being sold on America's Union Pacific Railroad.

By the mid-1990s, total production had reached to about 13 million hectoliters at ten breweries. Like many large companies, Bass became the United Kingdom's largest through acquisitions. Perhaps the largest of these was the Charrington Brewery in London which dated to the eighteenth century and had grown large

Above: The classic labels of the breweries who combined in 1960 to form Scottish & Newcastle. Scottish Brewers was formed in 1931, using the McEwan's brand name, which dates back to 1856. Newcastle Brown, the flagship of Newcastle Breweries, was created in 1927. Both brands still exist.

Opposite top: Historically, Bass was the biggest of the United Kingdom's Big Six. It's red triangle logo is Britain's oldest trademark.

Opposite bottom: The classic labels of the other four of the Big Six, Courage, Ind Coope, Whitbread, and Watney.

(These images from the author's collection were supplied by request by the breweries in the 1990s)

Above: Enjoying a pint of Morland's classic Old Speckled Hen. (Greene King)

Below: A summer day in the country with friends. (Greene King)

through acquisitions itself, becoming Charrington United in 1964 by taking over United Breweries. Bass and Charrington joined forces in 1967, and Charrington brought with it H&R Tennent, with the huge Wellpark Brewery in Glasgow, which it had acquired in 1963.

The turn of the century, however, brought change. The Bass hotel and pub activities had been spun off, and the brewing business was sold. The core Bass and Charrington brands went to Interbrew (AB InBev since 2008), although production of Bass was licensed to Marston's. Worthington brands were sold to Coors (later Molson Coors), and Tennent's brands (now Tennent Caledonian) went to the Ireland-based C&C Group (Cantrell & Cochrane), famous for Magner's Cider.

Bass brands brewed today include the flagship 5 percent ABV Bass Pale Ale, 6 percent India Pale Ale, and 3.8 percent Bass Extra

Smooth. Also produced is 3.6 percent Worthington Creamflow Bitter.

Ind Coope, which was the cornerstone of the Allied Group, was traditionally second only to Bass among Britain's largest brewing companies. The Star Inn, a brew-pub in Romford with an excellent reputation was purchased in 1779 by Edward Ind and acquired in 1845 by Octavius Edward Coope and his brother George. In 1856, Ind Coope made the move to Burton and was soon brewing pale ale and India pale ale, which they branded with their Double Diamond. In 1934, Ind Coope merged with Allsopp's.

In 1961, Allied Breweries was born through the merger of Ind Coope, Tetley Walker of Warrington, and Ansells Brewery of Birmingham. Allied, along with Carlsberg, was one of the breweries to launch the international lager brand, Skol. After the 1978 acquisition of food catering company J. Lyons, the company became Allied-Lyons in 1981. By 1992, Allied-Lyons was brewing 8.3 million hectoliters annually and owned 600 pubs and shops. In that same year, the Allied brewing activities were spun off and sold to Carlsberg, which created its Carlsberg-Tetley subsidiary. This entity sold the venerable Ind Coope Burton Brewery to Bass in 1997 and later wound up with Molson Coors.

Still brewed today are 4.8 percent ABV Ind Coope Burton Ale, 3.3 percent Tetley's English Ale, 3.6 percent Tetley's Original Bitter, 5.2 percent Tetley's Trust Porter, 4.2 percent Huntsman's Ale, and 2.8 percent Skol Lager.

The ancestors of Scottish & Newcastle, formed in 1960, go back at least to 1749, with the Scottish Breweries predecessor formed by the 1931 merger of McEwan's and William Younger & Company. Newcastle Breweries, meanwhile, was an amalgam of breweries that included the Royal Brewery and the Tyne Brewery, both in Newcastle. Their flagship Newcastle Brown Ale was launched in 1927. Scottish & Newcastle grew into one of the United Kingdom's largest operations, surviving the demise of rival Allied-Lyons by more than a decade. It acquired Kronenbourg in France and Hartwall in Finland. Scottish & Newcastle even briefly owned former rival Courage, and once controlled half of Baltic Beverages Holding (BBH), the huge holder of breweries across the former Soviet Union. Closer to home, they closed their own McEwan's Fountain Brewery near Edinburgh in 2004 and acquired the nearby Caledonian Brewery. Founded in 1869, Caledonian became the last survivor among the nineteenth-century Scottish breweries. In 2008, however, it was Scottish & Newcastle's turn

on the block. It was acquired jointly by Heineken and Carlsberg, who divided its holdings between them. Heineken took the United Kingdom assets of Scottish & Newcastle, spun off the pub properties, and renamed the business as Heineken UK. In 2011, Heineken sold the McEwan's brand to Wells & Young. The former Scottish & Newcastle beers distributed today include McEwan's 80 and McEwan's Export, both at 4.5 percent ABV, 4.7 percent McEwan's India Pale Ale, 8 percent McEwan's Scotch Ale, the iconic 4.7 percent Newcastle Brown, and 5.2 percent Newcastle Winter IPA.

**Above: At work in SABMiller's Brewing Research Facility, housed within the University of Nottingham's School of Biosciences. The company commissioned the facility with the aim of pioneering new developments in the science of brewing. The research is "intended to lead to advances in the sustainability and resource efficiency of beer production, and could contribute to the development of significant consumer benefits, such as enhanced shelf life." (SABMiller)**

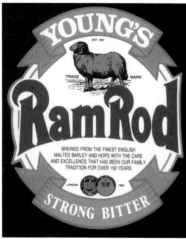

Above: The classic brands of Wells and Young's. Formed in 1876 and 1831 respectively, the two brewing companies merged in 2006. (Images from the breweries via the author's collection)

Samuel Whitbread began brewing on Chiswell Street in London in 1750, where he built the largest brewery in Britain. In 1869, Francis Manning-Needham is credited with having introduced the idea of bottled beer at Whitbread, and Whitbread had become one of the pioneers of packaged beer, along with Worthington and Bass. The company grew gradually for two centuries, then enjoyed a 350 percent growth spurt in the 1960s that took it to more than seven million hectoliters of annual output. The company-owned pubs, meanwhile, grew into a huge hotel and restaurant business that outperformed the brewing operations, so in 2001 the company sold its breweries to Interbrew (later AB InBev). Brewed today are 3.4 percent ABV Whitbread Best Bitter, 5.7 percent Whitbread Pale Ale, and 3 percent Mackeson Stout.

The Watney story goes back to 1837, when James Watney became a partner in London's Stag Brewery. In 1898, Watney's company merged with Combe, Delafield & Company (the Woodyard Brewery) and Reid & Company to form Watney, Combe & Reid. In 1958, this entity merged with Mann, Crossman & Paulin to form Watney Mann. During the middle twentieth century, the flagship ale, Watney's Red Barrel, was popular and widely visible in England and the export market. It even had the dubious distinction of being lampooned by Monty Python. Watney Mann was acquired in 1972 by Grand Metropolitan, a hotel and spirits conglomerate, and wound down. Red Barrel survived, albeit with gradually diminishing popularity, until 1997. Eight percent ABV Watney's Scotch Ale can still be found.

Courage & Company was founded at the Anchor Brewhouse in Southwark, London, by John Courage in 1787. Southwark brewing was already renowned. Chaucer noted its prowess in the fourteenth-century *Canterbury Tales*. Shakespeare based Falstaff on one of Southwark's leading brewing figures, Samuel Pepys viewed the Great Fire of London in 1666 from a Southwark ale house, and Dr. Samuel Johnson earned a few quid as an investor in the Thrale family brewery.

As with the other members of the Big Six, Courage grew steadily over the ensuing two centuries, both in terms of its beer sales and its pub and hospitality business. Interestingly, the origins of John Courage Amber Lager go back in years to a specially developed export beer that rapidly became the Navy's own brand. In fact, it eventually replaced the seaman's daily tot of rum when the Royal Navy ended that longstanding tradition. By the middle of the twentieth century, Courage operated five breweries in London and had additional breweries in Reading, Bristol,

Plymouth, and Newark, where it had acquired smaller brewing companies. In 1972, Courage itself was acquired by Imperial Tobacco, who closed the London brewery in 1981 and sold the company to Elders IXL (later Foster's) in 1986. The company was next sold to Scottish & Newcastle in 1995, which formed the subsidiary called Scottish Courage. When Heineken acquired the Scottish & Newcastle operations in 2008, Courage was divested, winding up as a subsidiary of Wells & Young's Brewing. They still market Courage Best Bitter, which is 4.0 percent in the cask and 3.8 percent in the bottle, as well as 10 percent Courage Imperial Russian Stout.

This, in turn, brings the beer traveler from the nostalgic tour of past centuries to the next generation of larger United Kingdom-owned brewing companies. Wells & Young's Brewery was formed in a 2006 merger of the brewing operations of Charles Wells Limited of Bedford and Young's Brewery in the London borough of Wandsworth. At the time of the merger, both remained family-owned.

Charles Wells Limited was founded by its namesake in

Above: Ale fermenters at Marston's in Wolverhampton. (Marston's)

Below: Highlights from the Fuller's product line in recently updated bottles and labels. (Fuller's)

**Above: Enjoying a pint at the Meantime Brewery in Greenwich.** (Meantime)

**Below: Organic Scotch Ale takes the foreground in this product shot at Black Isle Brewery near Munlochy, Scotland.** (Black Isle)

1876, when he retired from his first career as a sea captain. Following his death in 1914, his sons continued to develop the business by taking over other companies and their breweries and pubs. The Wells flagship is Wells Bombardier, which is 4.1 percent cask, 4.7 percent bottle. Also produced is 4.7 percent Wells Bombardier Burning Gold.

Young & Company's Ram Brewery evolved from a sixteenth-century brewery started in Wandsworth by Humphrey Langridge, which Charles Allen Young bought in 1831. When it closed in 2006, the London newspaper *The Independent* mourned the loss of "the oldest continuous beer-making site in Britain." Commented Peter Whitehead, the finance director at Young's,

"Brewing these days is all about scale in this very competitive market—it's scale that gives you profitability." Young's products include Young's Bitter, which is 3.7 percent in the cask and 4.5 percent in the bottle, as well as 5 percent Young's Ram Rod, 4.5 percent Young's Special, 4.5 percent Young's London Gold, 5 percent Young's Winter Warmer, 5.2 percent Young's Double Chocolate Stout, and 6.4 percent Young's Special London Ale.

With the departure of Young's, the last full-sized family-owned brewery in London was Fuller's (Fuller, Smith & Turner) Griffin Brewery in Chiswick. In 1829, John Fuller entered into a partnership in the brewery without knowing much about brewing, was abandoned by his partner, and was saved by his son. In 1845, John Bird Fuller brought in Henry Smith (late of the Romford Brewery) along with his brother-in-law, head brewer John Turner, and Fuller, Smith & Turner was born. The flagship ale is Fuller's London Pride, which is 4.1 percent ABV in the cask and 4.7 percent in the bottle. Other products include Fuller's India Pale Ale, which is 4.9 percent ABV in the cask and 5.3 percent in the bottle, as well as 5.4 percent Fuller's London Porter, 5.3 percent Fuller's Old Winter Ale, and 5.3 percent Past Masters Old Burton Extra.

As the old breweries go away, newer ones arrive continuously. A growing number of smaller and highly regarded microbreweries have recently appeared in London, one example being the Kernel Brewery, brewers of the popular 9.8 percent ABV Imperial Brown Stout London 1856. Kernel notes that their brewery "springs from the need to have more good beer. Beer deserving of a certain attention. Beer that forces you to confront and consider what you are drinking. Upfront hops, lingering bitterness, warming alcohols, bodies of malt. Lengths and depths of flavor. We make Pale Ales, India Pale Ales and old school London Porters and Stouts towards these ends. Bottled alive, to give them time to grow."

An example of an old British brewer that has grown through a vigorous expansion policy in recent years is Greene King, founded in 1799 in Bury St. Edmunds in Suffolk by Benjamin Greene, the great-grandfather of author Graham Greene. By the start of the twenty-first century, Greene King controlled around 500 local pubs and more than 200 destination pubs. Among Greene King's significant brewery acquisitions have been Morland in 1999, Ruddles in 2002, Ridley's and Belhaven in 2005, and Hardys & Hansons (the Kimberley Brewery) in 2006. The company also owns several chains of pubs and restaurants. Of the brewing companies acquired, only at Belhaven is the actual brewery still in operation.

Greene King products include 5 percent ABV Abbot Ale, 3.6 percent Greene King IPA, 3.6 percent Greene King IPA Draught Bitter, 5.4 percent Greene King IPA Reserve, 3.9 percent Hardys & Hansons Bitter, 6 percent Olde Suffolk English (Strong Suffolk Vintage) Ale, 3.7 percent Ruddles Best, 4.7 percent Ruddles County, and 4.4 percent St. Edmund's Gold. Belhaven beers include 3.9 percent ABV Belhaven Best Bitter, 5.2 percent Belhaven Scottish Ale, 4.2 percent Belhaven Scottish Stout, 4.2 percent Robert Burns Scottish Ale, and 6.5 percent Wee Heavy.

Established in 1711, Morland of Abingdon, Oxfordshire, was long heralded as England's second-oldest independent brewer after the Ram Brewery, with over 200 pubs, mostly in the Thames Valley

**Above: Specialty glassware at the Wellpark Brewery of Tennent Caledonian on Duke Street in Glasgow. (Tennent Caledonian)**

**Below: The Belhaven Brewery near Dunbar in Scotland. (Greene King)**

**Above: The product line of BrewDog, located in Ellon in Aberdeenshire. (BrewDog)**

**Below: Tactical Nuclear Penguin, with 32 percent ABV was one of the world's strongest beers (BrewDog)**

region. Morland was best known, perhaps, for its flagship ale, Old Speckled Hen, first brewed in 1979 to celebrate the 50th anniversary of the nearby MG sports car factory. Still produced by Greene King, it varies in ABV from 4.5 percent from the cask to 5.2 percent in the bottle. Also produced is the 6.5 percent Old Crafty Hen.

A near neighbor of Greene King in Suffolk, meanwhile, is a new-generation craft brewery called Old Chimneys that was started in 1995 in the village of Market Weston by master brewer Alan Thomson. Old Chimneys is highly regarded for their 11 percent ABV Good King Henry Special Reserve Russian Imperial Stout.

Venturing outside London at the south of England, the beer traveler will find an inexhaustible

number of pubs to enjoy, a myriad of beers and beer styles, and interesting experiences and fond memories of pubs and breweries from Cornwall to the top of Scotland.

Black Isle Brewery, on the isle of the same name in the Highlands of Scotland, is part of the new generation, dating to 1998. As they put it, they are "a long way from the standard industrial estate setting . . . nestled in the beautiful surroundings of the Scottish Highlands, a true beervana. We use only the best, freshest, organic ingredients for our brews. We are small, intensely independent and committed to the production of innovative and ground-breaking beer." Beers from Black Isle include 4.5 percent ABV Organic Blonde Lager, 5.6 percent Organic Goldeneye Pale Ale, 4.5 percent Organic Porter, 4.5 percent Organic Scotch Ale, and 4.5 percent Red Kite Ale.

BrewDog, Scotland's largest independent brewery, was founded by James Watt and Martin Dickie in 2007. With brewing in Ellon, Aberdeenshire, they operate pubs in Aberdeen and Edinburgh and distribute to many locations in Britain and abroad. Their portfolio includes 5 percent ABV 5am Saint, 9.2 percent Hardcore IPA, 5.6 percent Punk IPA, and 4.1 percent Trashy Blonde. Using freeze distillation to extract water, BrewDog has also produced some of the highest ABV beers ever made. These have included 32 percent ABV Tactical Nuclear Penguin, released in 2009; followed in 2010 by 41 percent Sink The Bismarck!, and 55 percent ABV The End of History. Bottles of the latter were packaged within stuffed animals, earning the ire of animal rights groups. BrewDog has been in a competition with the Schorschbräu brewery in Germany, where they have continued to top one another in the ABV race. Schorschbock is reportedly 57 percent ABV.

One of the most highly rated of Scottish brewers is the Harviestoun Brewery in Alva (Allamhagh Beag), a village in Clackmannanshire in the Central Lowlands. Ken Brooker started his brewery in 1984 in an eighteenth-century stone barn on the Harviestoun estate near Tillicoultry, but moved to Alva in 2004. The brewery was acquired by the Caledonian Brewery in 2006, but after the takeover of the latter by Scottish & Newcastle, Harviestoun became independent in 2008. Its most highly regarded ales are Ola Dubh Special Reserve 30 and 40, both at 8 percent ABV, and the 9 percent porter called Old Engine Oil Engineer's Reserve Blackest Ale.

Unusual within a beer culture with a long and well-established tradition of its own, the West Brewery and Bar brews German-style lagers. Started by Petra Wetzel

in 2006 in Glasgow, West was a German-style brewery and beer hall, but a new state-of-the-art brewery has been added in suburban Port Dundas. The flagship beer, named for Glasgow's patron

**Top: The Jennings Brewery in Lorton in the Lake District.** (Jennings)

**Above: Tap handles at the Cooperage Bar at the Jennings Brewery.** (Jennings)

Above: Tap handle clips bearing the iconic wild boar from the Ringwood Brewery at the edge of the New Forest in Hampshire. (Ringwood)

Below: A bottle of St. Mungo from Petra Wetzel's West Brewery in Glasgow. St. Mungo is the city's patron saint. (West)

saint, is St. Mungo, a 4.9 percent ABV lager described as "a hybrid between a true Bavarian Helles and a Northern German Pils, which makes for a complex yet very drinkable Premium lager."

The Jennings Brewery was established by John Jennings in 1828 in the village of Lorton in England's Lake District. Brewing relocated to Cockermouth in 1874. The company was acquired in 2005 by Martson's. Local fears that the Jennings Brewery would be closed have been thus far unfounded. Jennings ales with nineteenth-century roots include the 3.5 percent ABV Bitter and 4 percent Cumberland Ale. Two popular ales launched in the 1990s are 4.6 percent Cocker Hop and 5.1 percent Sneck Lifter.

Marston's, meanwhile, had evolved into the largest producer of cask ale in the world, and the landlord of more than 2,000 pubs. It dates back to 1834, when John Marston founded J. Marston & Son at the Horninglow Brewery in Burton. The company went through a long succession of merg-

ers and acquisitions, becoming the Wolverhampton & Dudley in 1890. The company readopted the Marston's name in 2007. Since 2005, the company has brewed the venerable Bass brand under license from AB InBev. In addition to Jennings and its own Burton facility, Marston's also operates the Park Brewery in Wolverhampton, the Ringwood Brewery in Hampshire, and the Wychwood Brewery in Witney.

Beers brewed by Marston's include Bank's Bitter, Burton Best Bitter, and Perfect Union, all 3.8 percent ABV bitters. Others include 3.6 percent Marston's EPA English Pale Ale, 5.5 percent Staffordshire IPA, and 6.5 percent Vasileostrovsky Imperial Russian Stout. Marston's markets a line of 4 percent ales under the "Single Hop" brand name, each of them made with a single hop variety. The varieties in the various beers include Cascade, East Kent Goldings, Marynka, Saaz, and Sovereign. Finally, Marston's brews "house brands" for British retail chains including Marks & Spencer, Sainsbury's, and Tesco.

The Ringwood Brewery was founded in 1978 near the fringes of the New Forest, where a number of breweries existed before 1923. As Michael Jackson wrote of Ringwood founder, Peter Austin, "While international lager-makers swallow one another up, Britain's real ale movement is still finding disciples in faraway lands [including Geary's Brewing and Shipyard in New England]. Peter Austin is their teacher. With his high forehead, silver hair and beard, Peter Austin has the look of a prophet. . . . He has three generations of English ale in his veins." Among the Ringwood ales are 3.8 percent ABV Best Bitter, 4.9 percent Fortyniner, and 5.1 percent Old Thumper, for which Austin was once awarded CAMRA's "Champion beer of Britain."

Yorkshire, the largest county in the United Kingdom by area, is the home to a pair of important and once-related brewing and pub owning companies named Smith. The origins of the two go back to the oldest brewery in Yorkshire, the Backhouse & Hartley Old Brewery, established in 1758 in Tadcaster. In 1847, Samuel Smith bought the Old Brewery from the Hartley family for his son John. He passed the Old Brewery to his nephew Samuel Smith, who

Above: The Marston's Brewery by night. (Marston's)

Below: The product line from Samuel Smith's Old Brewery in Tadcaster, Yorkshire's oldest brewery. (Merchant du Vin)

Above: London's Mayor Boris Johnson (left) and Meantime Brewery founder Alastair Hook at the brewery in 2010. (Meantime)

Below: Two beers from Sharp's in Rock, Cornwall. (Sharp's)

branded it with the current name in 1886. John, meanwhile started John Smith's Brewery. This brewery went on to evolve into one of the United Kingdom's largest brewing companies in the twentieth century and was acquired by Courage in 1970. In turn, Courage was acquired by Scottish & Newcastle in 1995, which became part of Heineken in 2008.

Samuel Smith's Old Brewery, meanwhile, remained as an independent brewing company, today operating around 200 pubs. Both Dr. Samuel Johnson and Oliver Goldsmith frequented Samuel Smith's Tadcaster pubs, and Charles Dickens describes the brewery in his *Tale of Two Cities*. Most of the beer produced at the brewery is fermented in "stone Yorkshire squares," fermenting vessels made of solid slabs of slate, using the same strain of yeast as in the nineteenth century. The beers of Samuel Smith's Old Brewery in Tadcaster include Samuel Smith's Famous Taddy Porter, Samuel Smith's Oatmeal Stout, Samuel Smith's Nut Brown Ale, and Samuel Smith's Organic Chocolate Stout, all rated at 5 percent ABV, as well as 7 percent Samuel Smith's Imperial Stout.

Also in Yorkshire is Rooster's Brewery, founded in Knaresborough in 1993 by Sean and Alison Franklin, and now run by the brothers Tom and Oliver Fozard. Their ales include 4.3 percent ABV American-inspired Yankee, 3.9 percent Wild Mule, and 4.3 percent YPA (Yorkshire Pale Ale), which has earned a World Beer Cup Gold Medal.

Sharp's Brewery was founded by Bill Sharp in 1994 at Rock, Cornwall, which evolved into the largest brewer of cask-conditioned beer in southwest England. In 2011, the brewery was sold to Molson Coors. In the meantime, Sharp's had established their flagship brand, 4 percent ABV Doom Bar Bitter, as one of the United Kingdom's leading cask ales. It is named for a sandbar at the mouth of the Camel Estuary in Cornwall. Other ales are 3.6 percent Cornish Coaster, named after a Cornish fishing boat, and 4.4 percent Sharp's Own, the first beer brewed at the brewery.

In Herefordshire, the beer traveler will encounter the ales of the Wye Valley Brewery, founded in 1985 by Peter Amor, a former Guinness brewer. Located in the village of Stoke Lacy since 2002, Wye Valley is managed by Peter's son Vernon, who had trained as a brewer at Young's in London. There are three flagship ales, 4 percent ABV Hereford Pale Ale, 3.7 percent Wye Valley Bitter, and 4.5 percent Butty Bach (meaning "Little Friend" in Welsh). Another line of ales, originally brewed as seasonal beers but now brewed all year, are branded under the name "Dorothy Goodbody." She is an imaginary person who is described in Wye Valley Brewery mythology as a hop grower's daughter. Among these are 4.2 percent Dorothy Goodbody's Golden Ale, 6 percent Dorothy Goodbody's Country Ale, and 4.6 percent Dorothy Goodbody's Wholesome Stout, which was named "Champion Winter Beer Of Britain" in 2002.

**Above: Chrome tap handles at Roster's Brewery in Knaresborough, Yorkshire. (Rooster's)**

**Below: The product line of Wye Valley Brewery in Herefordshire. (Wye Valley)**

# Scandinavia

In ancient times, the Scandinavians, like the Celts, drank beverages made with fermented grains because the climate was hardly favorable for the growing of grapes for wine. The development of their beverages influenced the Anglo Saxons, who in turn influenced—and were influenced by—the Celts. Countless small home and farmhouse breweries existed across Scandinavia for centuries before the first modern commercial breweries started to appear in the eighteenth century. Christmas beer, called Julöl (Yule ale), continued to be a popular beer style even after the big commercial brewers concentrated their production on lager in the nineteenth century.

However, Scandinavia, except Denmark, was also unique in Europe for adopting full-scale prohibition for periods of the twentieth century. As in the United States between 1920 and 1933, prohibition was introduced between 1914 and 1919. This was gradually lifted by 1932, but rationing remained in Sweden until 1955 and beer was banned in Iceland until 1989. Outside Denmark, alcoholic beverages are still highly taxed and heavily regulated.

Aside from the prices, the beer traveler will find two stories in Scandinavia: The first is about a sparsely populated region of Europe that, conversely, became one of the brewing world's great volume success stories. The other tale, equally important and probably more significant to the beer traveler, is a recent story of small

**Below: Enjoying a beer on a long, warm Scandinavian summer day. (Carlsberg)**

craft breweries that have percolated to the top in international listings of the world's best beers. Some of these breweries are Evil Twin Brewing in Copenhagen; Bryggeriet Djaevlebryg, also in Copenhagen; Haand Bryggeriet in Drammen, Norway; Jämtlands Bryggeri in Östersund, Sweden; Mikkeller in Copenhagen; Närke Kulturbryggeri in Örebro, Sweden; Nils Oscar in Nykoping, Sweden; and Nøgne Ø (aka Det Kompromissløse Bryggeri) in Grimstad, Norway.

In Scandinavia, we find the interesting circumstance that the region's leading brewing company actually produces more beer through its multi-national brewery holdings and licensing agreements than is produced by all of the other brewing companies in all five countries combined. The Carlsberg Group, based in Denmark, is roughly tied with Heineken as one of the four leading brewery groups in the world, these two being the only ones in the top four recognized by a single brand name. Carlsberg's flagship lager brand, Carlsberg, is one of the best-known beer brands in the world while this brand and the company's Baltika and Tuborg brands are among the top six brands in Europe. Annual worldwide production exceeds 120 million hectoliters.

Denmark, the largest brewing nation in Scandinavia, produces nearly seven million hectoliters annually, followed by Sweden and Finland with around five million each, Norway with about three mil-

**Above: Carlsberg founder Jacob Christian Jacobsen. (Carlsberg)**

**Below: The famous Elephant Gate at the Carlsberg Brewery in Copenhagen was designed by Vilhelm Dahlerup and built in 1901. (Carlsberg)**

**Above: Making a delivery to a retailer in Denmark.** (Carlsberg)

**Below: The flagship lagers of Carlsberg and Tuborg in the familiar bottles and cans, plus Tuborg's Black Lager. Denmark's two largest breweries merged in 1970 but still maintain distinct brand identities.** (Carlsberg)

lion, and Iceland with a bit more than 100,000 hectoliters.

Carlsberg as a brewing company was founded by Jacob Christian Jacobsen, who was born in 1811, the son of an already established Danish brewer, Christen Jacobsen of Copenhagen. The elder Jacobsen, aware of

Denmark's reputation, tried to teach his son a more scientific and systematic approach to the brewer's art by enrolling him at the newly established Copenhagen Technical University. In 1835, Christen Jacobsen died, leaving his 24-year-old son sole heir to one of Denmark's most advanced breweries. By this time, lager beer had suddenly captured the imagination of the world's beer drinkers, so Jacobsen made several trips to Germany in order to learn more about lager brewing. In 1847, Jacobsen began production in the Copenhagen suburb of Valby at his Carlsberg Brewery, named for his son Carl.

For many years Denmark's second biggest brewing company was Tuborg, which was founded in 1873 by Carl Frederik Tietgen at

Hellerup, north of Copenhagen. In 1970, as Denmark entered the European Common Market, the Big Two merged into an entity known as the Carlsberg Group. As was the case with Amstel in the Netherlands, when it joined the Heineken Group in 1968, Tuborg retained its own brand identity. Before the merger, Carlsberg and Tuborg had already planned on building their first breweries outside Denmark, Carlsberg in Malawi and Tuborg in Turkey. The first of many overseas brewing licenses went to Photos Photiades Breweries in Cyprus in 1966. Large additional breweries were built in the United Kingdom in 1974, Fredericia in Denmark in 1979, and Hong Kong in 1981.

In 2001, Carlsberg began acquiring ownership of Orkla, one of Norway's largest brewery holding companies, and in 2008 Carlsberg and Heineken jointly acquired Scottish & Newcastle, splitting its assets. In this transaction, Carlsberg got full control of Kronenbourg in France, the Greek brewery Mythos, and Baltic Beverages Holding (BBH). The latter was a holding company formed in 1991 by Pripps and Hartwall, the largest breweries respectively in Sweden and Finland, for the purpose of acquiring breweries in the former Soviet Union. In 2002, it had become a joint venture of Carlsberg and Scottish & Newcastle.

The Carlsberg Group brews a range of light-colored lagers, rated between 4 and 5.5 percent ABV, under its own name. These include Carlsberg Beer, Carlsberg Export,

**Above: The centerpiece of Royal Unibrew, now Denmark's second largest brewing company is the Faxe Bryggeri in Fakse. (Royal Unibrew)**

**Below: An aerial view, showing the Faxe Bryggeri as part of a Scandinavian winter wonderland. (Royal Unibrew)**

Above: The Aass Bryggeri in Drammen is the oldest brewery in Norway. (Aass)

Below: Aass Bock and Aass Gulløl, a pale lager. (Aass)

Carlsberg Gold, and Carlsberg Lite. Other notable products are 7.8 percent Carls Porter, 7.2 percent Elephant Beer (a lager), and a line of beers under the founder's name that include 5.5 percent ABV Jacobsen Extra Pilsner, 5.8 percent Jacobsen Dark Lager, and a 10.10 percent barleywine called Jacobsen Birthday Brew. Tuborg lager products include 4.6 percent ABV Tuborg Classic, 5.5 percent Tuborg Gold, 5.6 percent Tuborg Julebryg, and 4.8 percent Tuborg Schwarzbier, a black lager.

After the merger of Carlsberg and Tuborg, Denmark's second largest brewery was Ceres Bryggeriet, founded in Aarhus in 1856 by Malthe Conrad Lottrup, who named his company after the Roman goddess of grain and agricul-

ture. Two other Danish breweries of note were Albani, founded by Theodor Schiøtz in 1859 in Odense, and Faxe, founded by Nikoline and Conrad Nielsen in 1901 in the town of Fakse. Faxe was long allied with Thor Brewing, which had breweries in Hjorring, Randers, and Horsens. Between 1989 and 2000, these breweries joined forces under the same holding company called Bryggerigruppen, which became Royal Unibrew in 2005. Now Denmark's second largest brewing company, Royal Unibrew also has holdings in Finland, Latvia, Lithuania, and Poland and holds the license to brew Heineken for Denmark.

Ceres brands include 4.6 percent ABV Ceres Top Pilsner, 7.7 percent Ceres Royal Stout, 9.1 percent Ceres Oils 9, and 5.5 percent Ceres Havskum (Bering Bryg), a spiced beer. Albani's flagship brand has long been the 5.6 percent ABV lager Giraf, but also in the lineup is 7 percent Albani Julebryg, a Christmas beer. The Faxe portfolio includes 4.6 percent ABV Faxe Amber, 4.6 percent Faxe Fad, 5 percent Faxe Premium, and 10 percent Faxe Extra Strong. The Thor brands have been retired.

The oldest modern independent brewery in Norway started in 1843 as a trading company in Drammen that did a little brewing on the side. In 1860, the company was purchased by a farmer's son named Poul Lauritz Aass. (The surname rhymes with "Oz," not "Ass," contrary to what some comedians have suggested.) The business was lost in the great fire of 1866, but Aass reopened in 1867, and the company evolved into today's Aass Bryggeri.

Still family owned and managed, Aass takes pride in "satisfying our customers' needs better than our competitors . . . continuing the beer traditions and beer as a cultural product," adding that "we are extremely enthusiastic about innovation in all parts of the value chain. We are curious and innovative! We are un-snobbish, humble and love people."

Aass products include 4.7 percent ABV Aass Genuine Pilsner, 5.5 percent Aass Classic (amber), 6.5 percent Aass Bock, 5.5 percent Aass Gulløl, and 9 percent Aass Juleøl, a barleywine.

Another leading character in the history of Norwegian brewing in the nineteenth and twentieth century is the brewery founded in Oslo in 1876 by the brothers Amund and Ellef Ringnes. Their company grew into what was to be the largest brewing company in Norway. Ringnes sponsored the 1888–1896 polar expeditions by the great adventurer Fridtjof Nansen, and later those of Otto Sverdrup, who named islands in the Canadian Arctic after the Ringnes brothers.

Founded in Gothenburg by Johan Albrecht Pripp in 1828,

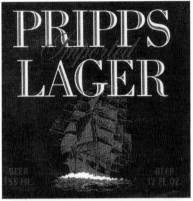

Pripps Bryggerier went on to be Sweden's largest brewing company with a roughly two-thirds market share, just as Ringnes held that position in Norway. As discussed above, with the dissolution of the Soviet Union in 1991, Pripps and Hartwall, Finland's largest brewing company, formed Baltic Beverages Holding to begin acquiring former Soviet breweries in the Baltic region.

By this time, both Ringnes and Pripps were part of the Orkla Group, a Norwegian food products conglomerate, and all were combined into a business unit that became a part of Carlsberg in 2004.

**Above:** The classic labels of the two great, venerable Scandinavian brewing companies, Pripps of Gothenburg, Sweden, and Ringnes of Oslo Norway. The two companies became part of the same business unit within the Orkla group in the 1990s, and have been owned by Denmark's Carlsberg since 2004. (These images from the author's collection were supplied by request by the breweries in the 1990s)

Above: Nøgne Ø brewer Ingrid Skistad on the bottle line. She came to brewing with a background in biotechnology and food science, and studied brewing and distilling at Heriot-Watt University in Edinburgh. (Nøgne Ø )

This ignited an enormous controversy in Norway, because for the first time, there was no Norwegian share in the ownership of Ringnes, Norway's largest brewery. Ringnes brands marketed today include 4.50 percent ABV Ringnes Ekte Fatøl and 4.6 percent Ringnes Classic as well as the 6.5 percent Frydenlund Bokkøl, a bock beer. The traditional Pripps products still in circulation include a range of lagers brewed in the 2.2 to 5.5 percent ABV range as well as 7 percent Pripps Extra Stark and 5 percent Pripps Julöl.

During the early twenty-first century, the craft brewing renaissance was in full swing across Denmark, Norway, and Sweden.

Nøgne Ø, founded in Grimstad in 2002, is Norway's leading and largest supplier of bottle-conditioned ale. The subtitle to the brewery name—Det Kompromissløse Bryggeri—means "The Uncompromising Brewery" and is described by the company as "a plain statement of our mission to craft ale of highest quality, personality and individuality."

The name Nøgne Ø itself means "naked island," a poetic term used by Henrik Ibsen to describe any of the countless stark, barren outcroppings that are visible in the rough sea off Norway's southern coast. The company statement adds, "We think the name also gives a symbolic picture of what the two Norwegian home brewers did early 2002, when they decided to found a new brewery. Their urge to share their passion of good beer was unbearable. Their vision was to bring diversity and innovation into commercial brewing of real ale."

Nøgne Ø year-round beers include 4.5 percent ABV Wit, 4.5 percent Brown Ale, 4.5 percent Bitter, 4.5 percent Havrestout (oatmeal stout), 6 percent Pale, 7.5 percent IPA, 10 percent #100 (barleywine), 7 percent Porter, 7.5 percent Imperial Brown, 9 percent Imperial Stout, 9 percent Tiger Tripel, 6 percent Brun (Belgian dark ale), 6.5 percent Saison, and 8.5 percent Two Captains (imperial IPA). There is also an extensive, though changing, portfolio of seasonal products.

Jämtlands Bryggeri in Östersund, Sweden, began brewing

**Above:** The Nøgne Ø brewery was founded in Grimstad, Norway in 2002. (Nøgne Ø )

**Opposite bottom:** The Nøgne Ø product line, left to right, includes Brun, Harvest Stout, IPA, Pale Ale and a Special Holiday Brew produced in collaboration with Jolly Pumpkin Artisan Ales of Dexter, Michigan and Stone Brewing of San Diego. (Nøgne Ø)

**Above: Nils Oscar Pils from Nykoping, Sweden.** (Nils Oscar)

**Below: Products from Jämtlands Bryggeri in Östersund, Sweden.** (Jämtlands)

at nearby Pilgrimstad—close to the place where pilgrims heading to the grave of Olav the Holy in Trondheim stopped to drink the pure water—and evolved as Sweden's most award-winning brewery. Products include 7 percent ABV Baltic Imperial Stout, 5 percent Bornsten (dark lager), 6 percent Black IPA, 5.5 percent IPA, 5.1 percent Hell, 4.5 percent Pilgrim (pale ale), 5.8 percent Postiljon (ESB), 5.2 percent President Pilsner, and 6.5 percent Julöl.

Nils Oscar Bryggeri in Nykoping, Sweden, started in Stockholm in 1996 as the Kungsholmens Kvartersbryggeri i Kungsholmen, and was renamed to the present name in 2006. Nils Oscar was a Swedish adventurer who, around a century earlier, made several arduous trips to the United States to work on farms and study agriculture. Though no mention is made of him having been a brewer, his diligence in his pursuits inspired this brewery's owners. Today, the brewery cultivates and malts its own grain for both brewing and distilling. Their bottle-conditioned products, "born on the fields and grown in the soil," include 4.7 percent ABV God [Good] Ale, 5.3 percent God Lager, 7.2 percent Nils Oscar Fyra Arstider Höst (farmhouse ale), 5 percent Nils Oscar Pils, and 10.4 percent Nils Oscar Barley Wine.

Mikkeller, founded in Copenhagen, Denmark, in 2006, was named for founders Mikkel Borg Bjergsø, who is still with the company, and Kristian Klarup Keller, who has since moved on. As they explained it, their objective was to "brew beers that challenge the

concept of good beer and move people. We do this by using the best ingredients and work with the most talented and creative minds around the world [and] make quality beers a serious alternative to wine and champagne when having gourmet food."

Though it was named "Danish Brewery of the Year" several times, Mikkeller has no traditional brewery and is described as a "phantom microbrewery" because its brewers brew at multiple locations. The nexus of Mikkeller activity is at its bars, of which there are several in Denmark, and one which opened in the United States in San Francisco in 2013.

In its first few years, Mikkeller brewed more than 600 different beers in a wide variety of styles, many with high ABV. A few examples include 7.5 percent ABV Beer Geek Breakfast, 10.9 percent Beer Geek Brunch Weasel, 17.5 percent Black, 8 percent It's Alive Grand Marnier Barrel Aged Wild Ale, 10 percent Monk's Elixir Quadrupel, 8.2 percent Spontan Cherry Frederiksdal Lambic, and 16 percent X Imperial Stout.

Another Copenhagen phantom brewery is Bryggeriet Djaevlebryg (Devil's Brew Brewery), whose slogan, "Satans gode øl," means "devilishly good beers." The company was started in 2006 by home-brewing brothers Rune and Stinus Lindgreen. Originally they brewed at the Brøckhouse in Hillerød, but since 2009 they have been at the Herslev Bryghus near Roskilde. The Djaevlebryg product line includes 6.5 percent ABV Dark Beast black ale, 6 percent Mareridt (Nightmare), 6.5 percent Mystikale spiced beer, 8.5 percent Nekron Stout, 6.5 percent Son of Nekron Porter, and 10.5 percent Pride of Nekron Imperial Stout.

**Below: As the Mikkeller team writes "the big breakthrough in the international beer world arose with the clean and simple idea of adding french press coffee to an oatmeal stout. The result 'Beer Geek Breakfast' was voted number one stout on the international beer forum Ratebeer.com and kick-started Mikkeller's international success." (Mikkeller)**

**Above: Pouring a glass of beer at the Nørrebro Bryghus, a Copenhagen brewpub. (Nørrebro)**

**Below: Some beers from the portfolio of Oy Sinebrychoff (aka "Koff") in Helsinki, Finland. (Sinebrychoff)**

In Finland, which was an autonomous part of Russia from 1809 to 1917, the three largest brewing companies all date back to the nineteenth century. In order of their founding, they are Oy Sinebrychoff (aka "Koff"), started in 1819; Oy Hartwall, started in 1836; and Olvi Oyj, which dates back to 1878. Of the three, Hartwell is the largest, and Olvi is the only one still independent.

The oldest brewing company in all of Scandinavia, Sinebrychoff was founded in Helsinki by the Russian merchant and distiller Nikolai Sinebrychoff. Owned by the family through most of the twentieth century, Sinebrychoff is now based in Kerava and part of Carlsberg. Lagers brewed under the flagship Koff brand range from 2.5 percent ABV Koff I to 5.2 percent Koff IVA. Other lagers

include 4.6 percent Karhu III, 5.3 percent Karhu IVA, and 6 percent Jouluolut IV. Also a longtime part of the lineup is 7.2 percent Sinebrychoff Porter.

Prior to Finland's entry into the European Union in 1995, beer was classified by ABV for tax purposes. From Class I, it ranged up to Class IVA (4.8 to 5.2 percent) and Class IVB (5.2 to 8.0 percent). The latter two classes were available only in restaurants and on the export market. Since 1995, the Roman numerals are retained by brewers only for nostalgic reasons.

Also based in Helsinki, Hartwall was founded by Victor Hartwall and largely family owned until it went public in 1994. Hartwell acquired another major Finnish brewer, Oy Mallasjuoma, founded in 1912 by Finnish manufacturer Henrik Mattsson and his

two sons, the brewer Ernst and the banker Max. After being taken over by Hartwell, the brewery's products were discontinued.

In 2002 Hartwall was acquired by Scottish & Newcastle. As noted above, Carlsberg and Heineken jointly acquired Scottish & Newcastle in 2008, splitting its assets. In this transaction, Carlsberg got full control of Baltic Beverages Holding (BBH), the 1991 joint venture of Hartwall and Pripps, while Heineken got full control of Hartwall. Products include 4.6 percent ABV Karjala III, 5.2 percent Karjala Export IV B, 6.3 percent Karjala Terva, 4.5 percent Lapin Kulta III, and 5 percent Lapin Kulta Premium Lager.

Olvi Oyj remains, to date, the largest independent brewing company in Finland. According to the company, it was founded in Iisalmi by master brewer William Gideon Aberg and his wife Onni "for the purpose of fighting drunkenness. In the spirit of Zacharias Topelius they wanted to offer milder alternatives to citizens possessed by a lust for spirits." In addition to the Finland brewery, Olvi owns breweries in Belarus and in each of the Baltic countries. Olvi products on the Finnish market, all lagers, include 4.5 percent ABV Olvi III Special, 4.7 percent Olvi Sandels, and 8.5 percent Olvi Tuplapukki IV B.

Panimoravintola Koulu is part of the new generation craft brewing movement in Finland. The two-story brewery restaurant and beer garden is located in a culinary school in the city of Turku. The beers brewed, all lagers, include 4.7 percent ABV Lehtori, 5.5 percent Maisteri, 4.7 percent Ope, and 7.2 percent Reksi, a bock beer.

In Iceland, a number of small brewers have cropped up since brewing was legalized in 1989. The Egill Skallagrimsson Brewery was founded in Reykjavik in 1913, two years before prohibition, and survived for 74 years manufacturing soft drinks. Modern lagers include 5.6 percent ABV Boli, 5 percent Gull, 5.6 percent Malt Jólabjór, 5.6 percent Porrabjór, and 6.2 percent Sterkur.

**Above: At the bar at Panimoravintola Koulu in Turku Finland.**
(Panimoravintola )

**Below: Inspecting the brew kettle at Panimoravintola Koulu.**
(Panimoravintola )

# The Baltic

**H**istorically squeezed between the rival political and cultural influences of Russia, Prussia, and Sweden, the Baltic states of Estonia, Latvia, and Lithuania borrowed from all three, although lager, borrowed from Germany, and porter, which originated in England, are predominant. Part of the Russian Empire, but heavily influenced by German culture through the nineteenth

century, the Baltics were occupied by Nazi Germany during World War II and fiefdoms of the Soviet empire until 1991. Since independence, nineteenth-century breweries which languished under Soviet domination—brewing the tasteless "state standard" beer known as Zhigulevskoye (aka Zhiguli or Zigulinis)—were refurbished, mainly through investments and acquisitions by established Western brewing companies.

The two biggest investors in the Baltic breweries have been Finland's Olvi Oyj and Baltic Beverages Holding (BBH), who today own at least a controlling interest in the region's largest brewers. Founded in 1991 by the Hartwall of Finland and the Pripps of Sweden, BBH was gradually taken over completely by Denmark's Carlsberg by 2008.

The Olvi portfolio includes A. Le Coq in Estonia, Cesu Alus in Latvia, Volfas Engelman (formerly Ragutis) in Lithuania and Lidskoe Pivo in Belarus. Carlsberg, meanwhile, controls or owns the

**Below: Bartenders at work in Place Beer Colors, the popular and trendy bar in Tartu, Estonia. (A. Le Coq)**

Saku Brewery in Estonia and the Aldaris Brewery in Latvia as well as Svyturys and Utenos in Lithuania. BBH also owns Baltika, the largest brewing concern in Russia.

In general, all of the Baltic area breweries are diversified beverage companies, producing and marketing not only beer, but a wide variety of other malt beverages and soft drinks as well as kvass, the traditional low-alcohol beverage made from rye bread. As in Germany and elsewhere, they also produce radler, a mixture of beer and a soft drink, usually carbonated lemonade.

A. LeCoq, the largest drinks manufacturer and the oldest continuously operating brewer in Estonia, traces its origins to a small brewery started in Tartu by B.J. Hesse in 1800, later owned by J.R. Schramm, and renamed as the Tivoli Brewery in 1893. Albert LeCoq, meanwhile, was a Prussian merchant of French Huguenot origin who started bottling Russian Imperial Stout in London for export to the Baltic countries in 1830. In 1906, the firm of A. LeCoq moved from London

to St. Petersburg intending to brew Russian Imperial Stout in Russia. However, they decided instead to brew in Estonia, acquiring the Tivoli Brewery in Tartu in 1912. Between 1921 and 1939, A. LeCoq grew into the largest brewery in the independent Republic of Estonia.

Beginning in 1940, under the Soviet occupation of Estonia, A. Le Coq was nationalized and renamed Tartu Olletehas. When the Germans occupied Estonia between 1941 and 1944, the brewery operated as Bierbräuerei Dorpat, supplying most of its output to the Wehrmacht. Between 1944 and 1991, Estonia was under Soviet occupation, and a nationalized A. LeCoq operated again as Tartu Olletehas, brewing state standard Zhigulevskoye beer for the

Above: Samples from the portfolio of the Bauskas Alus Brewery in Bauska, near Riga in Latvia. Seen here are Bauskas Gaisais (Bauska Light) and Bauskas Tumsais (Bauska Dark), and also offered are Marta Aalus (Märzen), Rigas Alus (Riga's Beer), and Sensu Alus (Ancestors' Beer). (Bauskas)

Above: Ekstra, a lager from Svyturys in Klaipeda, Lithuania. (Svyturys)

Below: Products from A. Le Coq in Tartu, Estonia. (A. Le Coq)

centralized Soviet economy. With the collapse of the Soviet Union and the restoration of Estonian independence, the brewery was privatized under the ownership of Magnum Konsuumer and Olvi. A holding company called A. LeCoq was formed in 1998, but that name was not reassigned to the brewery until 2007.

The Saku Brewery dates back to an estate brewery started in 1820 by the Prussian nobleman Karl Friedrich Rehbinder. At the time nearby Tallinn, the historical capital of Estonia, was known by its German name, "Reval," and Saku was known as "Sack bei Reval" meaning "Saku by [or near] Tallinn." The name would show up in brewery lore 180 years later.

The Rehbinder property was acquired in 1849 by Valerian Baggo, who expanded and modernized the brewery in Saku, creating his Saku Hele brand. The owners took the growing Saku Brewery public in 1899, by which time it had come to dominate the market in Tallinn. As with A. LeCoq, Saku flourished under Estonia's independence and suffered under Soviet domination until the union-wide efforts during the 1970s when the Soviet government sought to modernize its breweries. In 1991, with independence restored, the Saku Brewery was privatized and modernized through investments from BBH.

As the state standard Zhigulevskoye went away, the current Saku brand portfolio began

to take shape, with 4.6 percent ABV Saku Originaal first released in 1993. Next was the 5.3 percent ABV Saku Rock, described colorfully as "a beer with a tough character. It was first brewed in the spring of 1994 as a special edition beer for a music festival held that year. Nobody remembers much about the festival, but Rock is still alive. Good things always get noticed. Rock has character. Rock has power. Compared to other lager beers, Rock packs a stronger taste of hops, more punch and not a drop of mildness."

In 2000, to celebrate its 180th anniversary, Saku released a 5 percent ABV "super premium segment beer" under the historic name "Sack bei Reval." In 2004, Saku Originaal was introduced into the Finnish market, and in 2007 Saku started distributing Finnish Sinebrychoff products in Estonia.

While A. Le Coq and Saku both date to roots in the nineteenth century, Viru had its origins in 1975 during the Soviet brewery modernization era as the Viru Kolkhoz (Viru Collective). Privatized in 1991, it became Viru Olu. The 4.7 percent ABV Tooles lager was introduced in 1991, remaining in the Viru portfolio until 2010. In 1992, Viru released Joulumees (Santa Claus), establishing a tradition of annually releasing dark Christmas beers. In 2009, the brewery's Wiru Kadakaolu (Juniper Beer) was

named as "The Best Alcoholic Drink in Estonia." Beginning in 1997, Viru began producing the products of the Harboes Bryggeri in Denmark under license.

The Aldaris Brewery was founded in 1865 in Riga, Latvia, under the name Waldschlösschen (Forest Castle), at a time when German-speaking Riga was one of the most important industrial cities in the Russian Empire. Riga's place as an important Baltic seaport favored the marketing of the brewery's products. By the turn of the twentieth century, the brewery claimed to be Europe's most modern brewing facility. In 1937, when the Aldaris name was adopted, the brewery continued to keep pace technologically with new equipment. The name was retained during the decades of state control, and the brewery

Above: The bar Place Beer Colors in Tartu serves "beer cocktails" and tables have a button which is pressed to order a half pint of A. Le Coq. (A. Le Coq)

Below: In 2010, A. Le Coq received the Aurum Award from the European Oenogastronomic Brotherhoods Council (CEUCO) in recognition of facilitating the development and preservation of European regional food and drink culture. (A. Le Coq)

Above and below: Among the "treasure" located in 2011 on the wreck of the merchant ship *Oliva* were a number of beer bottles. These included the bottle of **A. Le Coq Russian Imperial Stout** which is seen being examined above. The *Oliva* went down in the Baltic Sea in 1869. (A. Le Coq)

survived a round of brewery closures in Latvia in the 1970s. When the brewery was privatized in 1992, it became one of many Baltic-area breweries in which Carlsberg invested, and in 2008 Carlsberg became the sole owner.

In addition to lager, porter, called porteris in Latvia, has been an important part of a lager-dominated Aldaris product line since the nineteenth century, and although it was abandoned in the early twentieth century, it returned in 1947. Marketed today at 6.8 percent ABV, Aldaris Porteris received a silver medal in the 2009 European Beer Star Awards. The brewery's product line also includes the 5.5 percent ABV Mezpils Alus, a lager originally created for the European Brewers Congress, held in Latvia in 1906.

The second-largest brewing operation in Latvia is Cesu Alus in Cesis, which is also the oldest brewery in the entire Baltic region and northeast Europe. The brewery traces its lineage to 1590 and the Cesis Castle's brewhouse. The commercial-scale brewery was opened in 1878 by a nobleman named Emanuel Sievers and went through a number of successor owners between 1922 and 1940, when it was taken over by the Soviet govern-

ment. Privatized in 1995, it was substantially upgraded with a new brewhouse opening in 2001.

Bauskas Alus Brewery in Bauska, Latvia, was established by the Soviet government in 1981. Its current product line is headed by Bauskas Gaisais (Bauska Light) and Bauskas Tumsais (Bauska Dark), and also includes Marta Alus (Märzen), Rigas Alus (Riga's Beer), and Sensu Alus (Ancestors' Beer).

Lithuania's two largest breweries in the late twentieth century were Utenos in Utena and Svyturys in Klaipeda. Gradually acquired by BBH after privatization in 1991, they were merged into a single entity called Svyturys-Utenos Alus in 2001. While Utenos opened in 1977 to brew Zigulinis as a Soviet state-owned brewery, Svyturys originated in 1784 when Klaipeda was part of Prussia and known as Memel. Started by J.W. Reincke, the facility became part of Theodor Preuss's Memeler Aktienbräuerei holdings in 1871. With Lithuanian independence in 1923, it became the Klaipedos Akcinus Alaus Bravoras. Under German occupation during World War II, it was the Memeler Ostquell Bräuerei, and after the war, under Soviet state control, it became the Svyturys (Lighthouse) Brewery.

Today, both the Utenos and Svyturys brands continue to be marketed, including 5 percent

ABV Utenos Auksinis and 6.6 percent ABV Utenos Porteris from Utenos. From the other half of the merged company come the flagship Svyturys Ekstra (5.2 percent ABV) as well as Svyturys Baltas, a 5 percent ABV hefeweizen; Svyturys Baltijos, a 5.8 percent ABV märzen; and 7 percent ABV Svyturys Adlerbock. Svyturys also pays tribute to its roots with its 5.3 percent ABV Memelbräu.

Another major Lithuanian brewery with German roots is Kalnapilis, which was founded in Panevezys in 1902 by Albert Foight. Originally called Bergschlösschen, the current name, which also means "small hilltop castle," was adopted in 1918. Acquired by BBH in 1994, it was sold to Denmark's Royal Unibrew in 2001. In turn, Kalnapilis was merged with the Tauras brewery in Vilnius as the Kalnapilio-Tauro Group.

Lithuania's third largest brewing company, Volfas Engelman, was founded in Kaunas in 1853 by Raphail Wolf (Volfas in Lithuanian), whose son later added the Ferdinand Engelman Brewery to the family's holdings, hence the present company name. By the 1930s, the brewery had grown to control about a third of the domestic market. In 1959, during the Soviet period, the brewery was combined with various other beverage producers into a collective based in and

„Memelbräu". A bit of Klaipèda region

named for the city of Kaunas, which was renamed Ragutis in 1967. After privatization and once again as Volfas Engelman, the brewery was controlled briefly by Pilsner Urquell before being acquired by Olvy in 1999. Volfas products include 6 percent ABV Imperial Porteris and 9 percent ABV Fortas Stipriausias.

Above: Memelbräu from the Svyturys brewery in Klaipeda, Lithuania commemorates the city's brewing tradition, dating back to the time when Klaipeda was located in Prussia and known as Memel. (Svyturys)

Below: The beer tent at the Ollsummer Festival in Tallinn, Estonia. Held in July, the festival started in 1993 and is Estonia's largest. (A. Le Coq)

# Russia

Russia is the third largest brewing nation in the world after China and the United States, with an annual output of more that 100 million hectoliters. When it was dissolved in 1991, the Soviet Union was in fifth place behind the United States, Germany, China, and Japan. Meanwhile, though, Russia ranks below twentieth by per capita consumption. This is due to the great popularity of vodka and the fact that any beverage with less than 10 percent ABV was officially defined as "food" until 2010!

One of the things that often surprises a first-time beer traveler in Russia is the use of plastic bottles. In the West, glass and aluminum packaging is preferred for beer, with plastic reserved for soft drinks. In Russia, beer is often sold in 1.5 to 2.5-liter (or larger) "PET" (polyethylene terephthalate) containers.

Traditional brewing in Russia goes back to the Middle Ages and includes kvass, a sweet, low-alcohol beverage mainly fermented from rye bread that is still made today. Popular throughout Eastern Europe, it is typically sold by vendors who roll it through the streets in barrels mounted on two-wheel carts.

In the seventeenth century, Tsar Peter the Great, himself a kvass drinker and a beer lover, encouraged large-scale barley production on the steppes south of Moscow, and the city of Voronezh became one of the most important

**Below: Studying the wort in the brewing hall at Baltika's St. Petersburg brewery. (Baltika)**

brewing centers in Russia south of the Baltic area. In the nineteenth century, brewing in Russia, as in most of Europe, continued to be done mainly by small local firms, although certain larger brewing companies appeared. One example was the brewery in Krasnoyarsk, established around 1875 by exiled Polish merchant Florian Klepacki. His Krasnoyarsk beer was popular and widely known throughout central Russia.

When the Russian Empire formally became the Soviet Union in 1922 after five years of revolution and civil war, private ownership of all businesses was prohibited, and the state took over. Gradually, consumer goods became standardized and were produced with less variety. Production of nearly all goods, including beer, was concentrated in large state-owned enterprises.

Essentially the same bland lagers were produced at numerous breweries throughout the Soviet Union and sold under a small number of "state standard" names.

The most common, almost universal, state standard "brand" was Zhigulevskoye, typically called Zhiguli and named for the Zhiguli Mountains near Samara. Other brands included Admiralteiskoye, Prazdnichnoye, Rizhskoye, and Slavyanskoye. A beer called Moskovskoye (Mockbckoe) was brewed in Moscow at Badajev Pivovarennij as well as at other state-owned breweries.

The Soviets built many breweries in the 1930s under the five-year plans, but just as the pre-revolution breweries had fallen into disrepair, the Stalin-era breweries were not modernized, and the state of brewing and the quality of beer in the Soviet Union was very poor. An effort was made in the 1970s to build a new generation of more modern breweries, but these projects took years to complete.

One notable variation on the oppressive Soviet-era stifling of innovation came in 1981, when brewers in Klin were permitted to create Klinskoye, an original brand, which became immediately

**Above: Baltika's Number 6 Black Porter is actually a black lager, not a classic porter. (Baltika)**

decade of the twenty-first century with 561 breweries. These included 40 large national brewers, 76 regional breweries, 263 microbreweries and 182 restaurant breweries. While the former two categories had remained relatively stable throughout the previous decade, the latter were on the increase and "new establishments with a brewery of their own are being opened all over Russia."

The largest brewing company in Russia in the twenty-first century is Pivzavod Baltika, or Baltika Brewery, with a domestic market share of around 40 percent. Headquartered in St. Petersburg, it is the second-largest brewing company in Europe. It also operates the Baltika Brew restaurant in St. Petersburg.

**Above: Baltika's flagship brewery in St. Petersburg. (Baltika)**

**Below: The Baltika products are mainly pale lagers, although Baltika 4 is an amber lager and Baltika 8 is a hefeweizen. (Baltika)**

popular with consumers in and around Moscow. Klinskoye is still widely available in Russia.

According to the Russian industry journal *Pivnoe Delo* (*Beer Business*), Russia began the second

The Baltika story is like the story of modern Russian brewing, as many of the major Russian breweries to emerge from Soviet state ownership have merged into the company since the turn of the twenty-first century. Baltika's flagship brewery originated as one of the breweries within the state-owned Leningrad Association of the Beer Brewing & Nonalcoholic Beverages Industry (Lenpivo). Construction began in 1978, but it was not completed until 1990. Though the brewery, still a state enterprise, was already known as the "Baltika Brewery," the beer was originally branded with state standard names. Privatization of the Baltika Brewery began in 1992, with the company reorganized as joint-stock company with more than 2,000 shareholders. In turn, the "Baltika" brand came into use.

In the 1990s, Baltika began investing in the restoration of three older, newly privatized breweries. The first was Klepacki's nineteenth-century brewery in Krasnoyarsk which had been bought out by former employees, who began brewing Kupecheskoye, a 4 percent ABV lager, in 1993. The other two were Soviet-era "Zhigulevskoye" breweries at Chelyabinsk and Yaroslavl, which dated to 1969 and 1974 respectively.

The Chelyabinsk brewery had gone through a number of name changes after having been privatized as Chelpiks in 1994. It

became Chelyabinskpivo in 1996 and Zolotoy Ural in 1999. In 2005, it was acquired by St. Petersburg–based Vena. The following year, the Baltika Breweries holding company acquired Pikra, Yarpivo, and Vena. Important regional brands such as Arsenalnoe, Chelyabinskoe, Uralsky Master, Yarpivo, and Zatecky Gus, have been retained as part of the larger Baltika portfolio.

In 1997, Baltika purchased a controlling interest in the recently privatized Donskoye Pivo brewery in Rostov-on-Don on the Black Sea. The former New Era Brewery had been built by the Soviet government in 1974. In addition to state standard brands, the plant had developed its own in-house products, such as Debut, Donskoye, Lux, and To Samoye. In 2000, Baltika acquired

**Above: Volga Amber in the classic Russian 1.5-liter "PET" plastic bottle. (Baltika)**

**Top: Working the bottle line at Baltika's Vena brewery. (Baltika)**

Above: Stainless steel kettles in the brewing hall at Baltika in St. Petersburg. (Baltika)

Below: On the left is Permskoye Gubernskoye from Perm, followed by a group of related brands including Klinskoye, Tolstiak, and two from Sibirskaya Korona. (Courtesy of the breweries)

the Tula Brewery at Tula, 120 miles south of Moscow, and opened Russia's largest malting facility in a joint venture with France's Groupe Soufflet.

As Baltika was acquiring, it too was being acquired. Through the years, Denmark's Carlsberg Group purchased an increasing interest in Baltika, and in 2012 Carlsberg completed its acquisition of the company. As it has for many years, Baltika continues to brew Carlsberg's brands for the Russian market.

Since the beginning of the twenty-first century, Baltika had built a number of new breweries throughout Russia. The largest of the new breweries is Baltika-Samara, serving the vast Volga region. It began producing in 2003, with production exceeding six million hectoliters annually. The million-hectoliter Baltika-Khabarovsk brewery, also opened in 2003, is the largest brewery in the Russian Far East. The Baltika-Novosibirsk brewery opened in 2007, with production intended to be 2 million hectoliters, but it was stepped up to 4.5 million hectoliters annually.

The portfolio today, all lagers, includes Baltika 1 through 3, pale lagers in the 4.4 to 4.8 percent ABV range; 5.6 percent Baltika 4 amber lager; a 7 percent black lager called Baltika 6 Black Porter; 5 percent Baltika 8 hefeweizen; and 8 percent Baltika 9.

Other brands from the Moscow region that are now part of the same group as Klinskoye include Tolstiak and the Sibirskaya Korona (Siberian Crown) family. First introduced in 1996 by the Rosar Brewery in Omsk, the Korona beers are now also brewed in the breweries of Ivanovo, Klinskoye, Perm, Saransk, and Volzhsky. In addition to the Sibirskaya Korona Klassicheskoye (Classic) and Sibirskaya Korona Originalnoe (Original), there

is a hoppy Sibirskaya Korona Prazdnichnoe, an amber Sibirskaya Korona Rubinovoye, and Sibirskaya Korona Beloye (White), an unfiltered wheat beer. These beers have been medal winners at several Sochi International Beer Festivals.

Another group of award-winners are the Permskoye Gubernskoye beers brewed in the city of Perm in the Ural Mountains. Also from Perm is a Pilsner-style beer that was created with input from Czech consultants, which is called Rifey after the ancient Greek name for the Urals. In its advertising, Rifey is dubiously claimed to be based on Soviet-era Zhigulevskoye. The product line includes the award-winning Rifey Uralskoye, with 4.9 percent ABV, and Rifey Osoboye Krepkoye, with an ABV of 7 percent.

Pikur, meanwhile, is a pale, traditionally brewed beer from Kursk. A line of beers of higher hop content than typical of Russian beers was introduced in Omsk in 1994 by Bagbier. These include the 7 percent ABV Bagbier Krepkoye.

**Above:** According to the Carlsberg website, the BBH brand Arsenalnoe is "a beer with a male character and is brewed for real men who value honour and strength." (Carlsberg)

**Below:** A Tolstiak poster proclaims "We Wanted Guys!" (Tolstiak)

# Poland

Poland is the fourth-leading brewing nation in Europe behind Russia, Germany, and the United Kingdom, with an annual output of nearly 40 million hectoliters of "piwo" (the Polish word for beer), more than double its production during the 1990s. Poland boasts more than two dozen microbreweries as well as more than 40 large commercial breweries, the largest of which have become grouped into one of three holding companies.

The largest brewery group is Kompania Piwowarska, with a 45 percent market share, followed by Grupa Zywiec, which has a 35 percent slice of the market. The controlling interest in both of the two groups is foreign-owned, by SABMiller and Heineken respectively. In turn, Carlsberg's Carlsberg Polska subsidiary has 14 percent of the Polish market, mainly through its ownership of the Okocim Brewery. In addition to brands which are specific to each of the breweries, the brewery groups controlled by the international conglomerates also brew the flagship lagers of those parent companies for the Polish market.

The brewing entities within Piwowarska include Tyskie, Lech and Dojlidy. The breweries of Tyskie Gorny Slask located in Tychy, in southern Poland near the Czech border, include one of the largest Soviet-era breweries in the country. Records show that Tychy's Fürstliche Bräuerei, or princely brewery, was operating early in the seventeenth century and that brew-

**Below: Pouring a Lech lager from a sleek, modern tap at a bar in Poland.** (Kompania Piwowarska)

ing has been ongoing ever since. In the twentieth century, the Tyskie produced porter as well as lager, but the lager brands are dominant today. The 5.6 percent ABV Tyskie Gronie and 5.7 percent ABV Tyskie Ksiazece lagers are exported, while at home, they account for about a fifth of all beer sold in Poland. Also brewed is 7 percent Debowe Mocne.

One of the largest breweries in modern Poland, Lech Browary Wielkopolski in Poznan (formerly the German city of Posen), boasts a capacity of 7.5 million hectoliters, vastly more than the mere 1.4 million hectoliters in annual production that it was brewing in the 1990s. The current brewery was opened in 1980, with a malting facility added in 1984. Since the operation was privatized in 1993, there have been numerous upgrades and expansions. The first new owner after Lech came out of stare control was the businessman Jan Kulczyk, who sold it to South African Breweries in 1996. In turn, SAB and Kulczyk jointly acquired the Tyskie Brewery,

and in 1999 operations of the two were combined under the Piwowarska umbrella.

In the late twentieth century, the brewery's portfolio included brands such as Bernardynskie, Jelen, Jubileuszowe, Kozlak (bock beer), Krotosz, Maltanskie, Ostrowski, Poznanski, Ratuszowe, and Regent as well as Lech. Today, the principal Lech products are a trio of light-colored lagers, Lech Premium (5.2 percent ABV), Lech Pils (5.5 percent ABV) and Lech Mocny, meaning Lech Strong, which is rated at 6.2 percent ABV. Other products include the 7 percent ABV Dobowe Mocne, meaning Strong Oak.

Located in the Dojlidy district on the south side of the city of Bialystok, the Dojlidy Brewery was started in 1891. It is often linked to the first brewery in Dojlidy,

**Above: Friends toasting with Tyskie at a sidewalk cafe on a warm summer evening. (Kompania Piwowarska)**

**Below: Bottles of Zubr (Bison) on the production line at Kompania Piwowarska' Lech Browary Wielkopolski in Poznan. (SABMiller/ One Red Eye)**

Above: The Zywiec brewery in Zywiec. (Zywiec)

Below: A selection of Polish beers, including lager and porter from Zywiec, Mastne from the Brackie Browar Zamkowy (Friar Castle Brewery), Kujawiak pilsner from the former Browery Bygoskie in Bydgoszcz, and Kaper lager from the Elbrewery in Elblag. (Zywiec)

which dates back to 1768 but was not active for much of the nineteenth century. The current brewery was built on the grounds of the Kruzensternow manor, then owned by Count Rüdigerow. In 1919, after World War I and Poland's independence, the Polish government gave the go-ahead

to restore the brewery, which had been stripped by retreating Russian troops. As the absentee heiress to the Kruzensternow estate did not have Polish citizenship, the government ordered the property sold.

The brewery was acquired and extensively rebuilt and upgraded by the Lubomirski family of Polish nobility, who at the time included Prince Stefan Lubomirski, the father of the Polish Olympic committee. The brewery flourished during the interwar period and was pressed into service to supply the needs of German troops during World War II. Destroyed in 1944 as Soviet troops began to occupy Poland, the brewery was again rebuilt, finally reopening in 1954 as the Bialostockie Zaklady Piwowarsko-Slodownicze. In 1996, the operation was privatized and sold to

Germany's Binding Bräuerei. In 2003, now renamed as the Dojlidy Brewery, it was acquired by the Kompania Piwowarska.

The flagship brand at Dojlidy is a 6 percent ABV lager called Zubr (Bison). Naturally, a European bison figures prominently on the label and in advertising. So too does the 1768 date, although neither the brand nor the brewery existed then.

The Grupa Zywiec, meanwhile, owns a dozen breweries operated by five subsidiaries, Cieszyn, Elbrewery, Lezajsk, Warka, and the Zywiec Brewery itself. The latter was founded in the southern Polish city of the same name, about 40 miles southwest of Krakow. Brewing began here in 1856, a date which is still printed on Zywiec labels. While some breweries have been owned by princes and noblemen, Zywiec was one of the breweries founded by Archduke Albrecht Frederick Hapsburg. The House of Hapsburg, which has included the monarchs of more than a dozen countries, as well as Holy Roman Emperors, ruled the empire of Austria and Austria-Hungary (which included Zywiec) from 1282 to 1918. Though the beer had a royal pedigree, the traditional Zywiec iconography, used in both labels and advertising, uses a theme from Polish folklore, a couple in traditional costumes dancing the Krakowiak, the old folk dance of Krakow and environs.

In the nineteenth century, Zywiec established and maintained a position as the largest brewery in southern Poland. During World War II, Zywiec was taken over by the Germans and operated as the Beskideribräueri Saybusch, under the assumption that the German ownership was permanent. When Poland was liberated in 1945, however, operations resumed as before the war. It was soon nationalized by the Polish communist government and operated as part of a centralized multi-brewery collective called Zaklady Piwowarskie.

After Zywiec was privatized in 1991, the Hapsburgs sued the Polish government over the confiscation, and the parties settled out of court for an undisclosed sum in 2005. In the meantime, Heineken acquired

**Above and below: In 2010, more than 100 brides visited the Lech Browary Wielkopolski in Poznan as part of a fundraiser called "Brides conquering the city for a sick child," which was on behalf of a five-year old boy named Maciek, suffering from epilepsy and autism, as well as heart and kidney defects. (Kompania Piwowarska)**

**Above: Serving beer at the At Grande Grill, just off the Main Market Square in Krakow. (SABMiller/One Red Eye)**

**Below: Beers from the Kompania Piwowarska breweries include Zubr (Bison) and Lech, as well as three from the Tyskie family, Debowe Mocne, and Ksiazece lager and märzen. (Piwowarskie)**

control and heavily modernized the facility. Current products from what is now referred to as the Zywiec Archduke Brewery include a 5.6 percent ABV lager called Zywiec Beer, 6.5 percent ABV Zywiec Bock, a hefty 9.5 percent ABV Zywiec Porter, and 5 percent ABV Brackie Pale Ale Belgijskie.

The nearby Cieszyn Brewery, also part of Grupa Zywiec, was also established by Archduke Albrecht Frederick Hapsburg, although it actually preceded Zywiec by a decade. Also known as the Brackie Browar Zamkowy (Friar Castle Brewery), Cieszyn brews the 5 percent ABV Brackie Pale Ale Belgijskie as well as the Zywiec brands, while Brackie is also brewed in Zywiec.

The Elbrewery, known as "EB," is the largest brewery in Grupa Zywiec, with a two million hectoliter capacity. Located in Elblag on the Baltic coast about 35 miles southeast of Gdansk, the first license was granted in 1309 to brewer Siegfried von Leuchtwangen, a master of the Order of Teutonic Knights. Located within Prussia and Germany for most of the years until World War I, the Elblag Brewery became the exclusive purveyor of beer to the German Kaiser. For many years, beginning in 1993, the flagship brand was the now retired EB Specjal. Current brands include two strong lagers, the 8.70 percent ABV Hevelius Kaper, and the 6.70 percent ABV Specjal Mocny.

The brewer's license for Grupa Zywiec's Lezajsk Brewery in the southeastern Poland town of the

same name was granted in 1525 by Polish King Sigismund I the Old. This fact is memorialized in the brand name of the 5.5 percent ABV Lezajsk 1525 Legendary Premium Lager.

Rounding out our tour of the Grupa Zywiec operations is the Warka Brewery in the town of the same name about 30 miles south of Warsaw. The company makes note of the fact that a brewery in Warka was named as the official supplier of beer to the court of Boleslaw V in 1478, but the present brewery was built by the Polish government in 1975 and sold to Zywiec in 1999. A 2004 modernization increased annual capacity to more that three million hectoliters. Among other products, Warka produces its 6.5 percent ABV Warka Strong Dwuslodowy.

The flagship brewery of the Carlsberg Polska Group is Okocim in Brzesko, a city in southeastern Poland. Okocim was founded in 1845 by a German brewer from Wirtemberg named Johann Goetz. The Carlsberg connection goes back to 1884, when Goetz hosted a visit from Carlsberg founder Jacob Christian Jacobsen. It would be 112 years and many decades of ups, downs, significant growth in market share, state confiscation, and state ownership before the two breweries came under the same ownership. The brands include 5.9 percent ABV Okocim (Zagloba) lager, 5.8 percent ABV Okocim Jasne Pelne lager, and 8.1 percent ABV Okocim Porter.

Other breweries in the Carlsberg Polska Group include Ciechan in Ciechanow and Lwowek in Lwowek Slaski. Brands include 9 percent ABV Ciechan Porter and 5.4 percent ABV lager Lwowek Ksiazecy.

Finally, we turn to a small but fast-growing brewing consortium that is actually based in Poland. Van Pur was founded in Warsaw in 1989 as a Polish-German joint venture. After buying out the Polish interests of the Danish consortium Royal Unibrew at the end of 2010, Van Pur had a total annual capacity of 3.5 million hectoliters in five breweries located in Jedrzejow, Koszalin, Lomza, Rakszawa, and Zabrze. Of these the brewery in Lomza is the largest, producing a range of lagers in the 5 to 6 percent ABV range as well as 7.8 percent ABV Lomza Mocne.

**Above: The brewing hall at Browary Tyskie Górny in Tychy. The brewery was founded in 1629.** (Kompania Piwowarska)

**Below: Checking the brewing kettles at the Lech Browary Wielkopolski in Poznan.** (SABMiller/One Red Eye)

# Eastern Europe

s we have seen in our armchair visits to Russia, Poland, and the Baltic, the story of modern commercial brewing across Eastern Europe begins in the nineteenth century with brewing operations often founded, or at least licensed, by the nobility, then fades into a twentieth century under the state control with little attention to quality. World War II and the decades of Soviet domination greatly diluted the region's brewing heritage while at the same time stifling innovation. By the turn of the twenty-first century, however, the surviving breweries in Eastern Europe have emerged from decades of neglect to be upgraded as some of the most modern in the world, thanks to investment from multinational conglomerates, especially Carlsberg and Heineken.

Interbrew (AB InBev after 2004) also made a substantial investment in Eastern Europe, but in 2009 sold many of its interests in the region to the London-based private equity fund CVC Capital Partners. CVC organized their holdings into an entity called StarBev, named after their flagship operation, the Staropramen Brewery (Pivovary Staropramen), which is the second largest brewing company in the Czech Republic. In 2012, CVC sold StarBev to Molson Coors.

It would be an accurate generalization to say that throughout the region the beer traveler will find these modern breweries serving international lagers as well as radler (the beer and lemonade blend that

**Below: Serving the products of Romania's Ursus Breweries. Sales of their Timisoreana brand have grown in double digits in the twenty-first century. (Jason Alden/One Red Eye)**

is popular in central and Eastern Europe), to a generation of consumers with no memory of the twentieth-century doldrums.

In Belarus, Carlsberg has a major presence with its Alivaria brand, while Heineken's Rechitsapivo is widely encountered by the traveler. However, the biggest selling beer, with more than a third of the market, is from a state-owned brewery, Brestskiy Kombinat Bezalkogolnych Napitkov in the city of Brest. The products include a 4.2 percent ABV dark lager called Brestskoe Temnoe and a 3.7 percent ABV light lager called Brestskoe Svetloe.

Bulgaria, where wine was the ancient beverage of choice, came to brewing late in the nineteenth century. Its first commercial brewery was probably the one established in 1876 by Swiss brewers Rudolf Frick and Friedrich Sulzer in the old Roman city of Plovdiv, now Bulgaria's second largest city. Another brewery was started in 1882 on Kamenitza Hill, near Plovdiv, which evolved into the Kamenitza Brewery of today. Now the market leader in Bulgaria, Kamenitza has gone through a series of owners in recent years as part of StarBev.

Among Bulgaria's other breweries, Pirinsko and Shumensko are part of the Carlsberg Group, while Ariana, Stolichno, and Zagorka are

**Above: A brewery worker with a case of Timisoreana at the Ursus brewery at Cluj Napoca in Romania. (Jason Alden/One Red Eye)**

Above: Checking the brew at the Ursus brewery. (Jason Alden/One Red Eye)

Below: Soproni line includes several lagers, such as Arany Facan (Golden Pheasant) and Meggy (Sour Cherry) Demon, a dark lager. (Heineken Hungaria)

Heineken brands brewed at the Zagorka Brewery in the city of Stara Zagora. In 2011, Heineken launched the 6 percent ABV Zagorka Rezerva, described as "the first ever winter drink that combines the best of beer and elements

typical for the good red wines—elegant aftertaste and pleasant fruity aroma." Molson Coors's StarBev, meanwhile, owns Kamenitza as well as the Astika, Burgasko, Plevensko, and Slavena labels.

An independent Bulgarian brewery is Bolyarka (formerly Velikotarnovsko) in the city of Veliko Tarnovo, which dates back to 1892, when the company was founded by the Hadji Slavchevi brothers. The product portfolio includes the 4.1 percent ABV Bolyarka Light, the 5 percent ABV Bolyarka Dark, and Bolyarka Weiss at 5.4 percent ABV.

Hungary's modern brewing tradition evolved during the nineteenth century at a time when the country was part of the Austro-Hungarian Empire, whose capital

and cultural nexus was Vienna, Austria. Hungary's first commercial brewery is said to have been started in 1845 in the Buda half of Budapest by Munich-trained brewer Peter Schmidt. In 1862, the famous Vienna lager brewer, Anton Dreher, bought a struggling brewery in the Kobanya district of Buda and turned it into what was to be the largest in the country, with a market share of 70 percent by 1933, until state control was imposed on Hungary after World War II. Today, the Dreher Brewery (Dreher Sorgyarak) is owned by SABMiller, which acquired rights to the Dreher name in Hungary in 1997. In Italy, where Anton Dreher started a brewery in Trieste when it was part of the Austro-Hungarian Empire, the name is owned by Heineken.

Another major empire-era Hungarian brewery traces its roots back to the Hirschfeld Sorgyar Reszveytasasa, founded in 1848 by Leopold Hirschfeld in Pecs in southwestern Hungary. In 1907, Leopold's son and heir, Samuel Hirschfeld, created the Szalon Sor (Salon Series) brand, a 4.6 percent ABV lager that is still the brewery's flagship brand. Now known as Pecs Brewery (Pecsi Sorfozde), the brewery is partially owned by Austria's Ottakringer Group but operates independently.

Two of today's largest modern Hungarian breweries were built under state control by the Magyar Orszagos Soripari Troszt

**Above: Pouring mugs of Timisoreana lager at a pub in Romania. (Jason Alden/One Red Eye)**

(Hungarian National Brewery Trust). Borsodi Sorgyar in Bocs in northeastern Hungary is now a StarBev brewery, and the former state brewery in Martfu (Martfui Sorgyar) is now part of Heineken Hungaria.

The latter entity was formed as a Heineken subsidiary in 2004 to manage the breweries purchased from Austria-based Bräu Union, which had acquired them

**Above: The slogan suggests that the end of the work day is time for a glass of Topvar.** (Topvar)

**Below: From Ukraine we see Obolon Premium Lager, as well as Rogan's Monasyrske Bright, Non-Alcoholic Light , Traditional Bright, and 6.9 percent ABV Veseliy Monakh Strong.** (Courtesy of the breweries)

cent ABV Munich-style dark lager called Soproni Fekete Demon.

In Romania, the fastest-growing brewing company is Ursus (Compania de Bere Romania) in Cluj Napoca, which dates back to 1878 and which was acquired by South African Breweries in 1997. Sales of its popular 5 percent ABV Timisoreana lager experienced double-digit annual growth early in the twenty-first century.

Serbia, which is the leading brewing nation of the former Yugoslavia, is home to a number of sizable brewing companies, the largest of which is Apatinska Pivara in Apatin, with an annual production of 4.5 million hectoliters. Founded in 1759, it went through a cycle of state control in the twentieth century and became part of the StarBev Group.

after privatization in the 1990s. Also part of Heineken Hungaria is Soproni Sorgyar, which was started in 1895 in the city of Sopron on the Austrian border. Today, the Heineken Hungaria products include the 4 percent ABV lager called Arany Facan (Golden Pheasant) and a 5.5 per-

Displaced by Apatinska as the largest Serbian brewery at the turn of the twenty-first century, Beogradska Industrija Piva, or BiP (Belgrade Industry Of Beer), is located in Belgrade, the national capital. In 2007, it became the last major brewery to be privatized.

In 1892, Lazar Dunderski founded Pivara Celarevo on a farm near the town of Celarevo in Serbia, and on this farm grew both hops (four varieties) and barley, which guaranteed a steady supply of essential ingredients. Technically upgraded in the 1970s as one of the most modern breweries in what was then Yugoslavia, the facility is now the second largest in Serbia, with a quarter of the total market. Controlling interest was acquired by Carlsberg in 2003 under the name Carlsberg Srbija.

Heineken's presence in Serbia came through its 2008 acquisition of Novosadska Pivara in Novi Sad. Also called Pivara MB, it was built in 2003 by the Serbia-based investment firm of Rodic M&B, under the slogan "Svetsko, a nase" ("Global, but ours"). It was "ours" for less than five years. In 2008, Heineken also acquired breweries in Pancevo and Zajecar, which had been purchased in 2004 by Turkey's Efes Group. Heineken then formed a company known as Ujedinjene Srpske Pivare (United Breweries of Serbia) which uses the MB brand name.

Slovakia, the eastern part of the former Czechoslovakia, has a small brewing industry which is dwarfed by the giants of the Czech Republic. SABMiller's Slovak holdings include two formerly independent lager lines which, since 2010, have both been brewed in Velky Saris at the brewery of Pivovar Saris. The other brewery, Pivovary Topvar, originally brewed in a state-owned plant that opened in Topolcany in 1964.

Heineken Slovakia, meanwhile, has a controlling share in the KK Company Brewery in Martin and the Gemer Brewery in Rimavska Sobota. Heineken's flagship brand is the 5 percent ABV Zlaty Bazant (Golden Pheasant), which is brewed at Pivovar Zlaty Bazant in Hurbanovo, which originated in 1967 as a state plant. It will be noted that in neighboring Hungary, Heineken Hungaria brews a lager called Arany Facan, which also translates as Golden Pheasant.

Ukraine, the largest country in Europe by area outside Russia, has long shared a brewing tradition with that country, especially while they were both united within the Soviet Union. However, just as Ukrainians have taken great pride in their independence,

Above: Bringing beer to the vineyard, she serves up a tray of Topvar on a warm September day. (Topvar)

Above: A BiP rep discusses her beer with a consumer at the Belgrade Beer Festival. (Beogradska Industrija Piva)

Below: White Night spiced beer, Maximum Beer (domestic and export), Pub Lager, and Yantar lager. (Chernihivske)

they have maintained independent, indigenous beer brands. While dominated by lagers, the product mix at Ukrainian breweries also includes Baltic porter and other products from the traditional kvass to the popular radler.

The principal brewing operations in Ukraine include one major home-grown company and three multinationals. The Ukrainian-owned company is the Obolon Joint Stock Company, whose flagship brewery in the capital city of Kiev first opened as a state plant in 1980, having been designed by Czech engineers. In 1992, when Ukraine became independent, Obolon was the first company to be privatized. Financed by loans from the European Bank for Reconstruction and Development, the Obolon facility in Kiev has been expanded into Europe's largest-capacity single brewing plant, and its malting facility in the Khmelnytskyi Oblast also has the largest capacity in Europe. The company also operates additional

breweries in Ukraine at Bershad, Chemerivtsi, Fastiv, Krasyliv, Kolomyia, Oleksandria, Okhtyrka, Rokytne, and Sevastopol. Obolon products include 5.2 percent ABV Obolon Premium Lager and 7 percent ABV Obolon Porter.

The flagship lager brand for the former SUN Group, now SUN InBev, is Chernihivske, originally brewed in Chernihiv, but now produced at plants in Mykolaiv and Kharkiv. Product variants include 5.6 percent ABV Chernihivske Premium, 7.5 percent ABV Chernihivske Strong, and 5 percent ABV Chernihivske White, which is said to be the first unfiltered beer on the Ukrainian market. SUN InBev also produces the Yantar and Rohan brands.

SUN is currently the largest of the international brewers in Ukraine, with a better than one-third market share in the country. The smallest international player is SABMiller, with its Sarmat brand.

Carlsberg, which also owns Baltika, Russia's largest brewing entity, operates several Ukrainian brewing operations. The Slavutich Brewery at Zaporizhia was acquired in 1996, and a new brewery in Kiev opened in 2004. The flagship lager is the 4.3 percent ABV Slavutich Svitle, but the breweries also produce international Carlsberg and Tuborg lagers as well as Baltika for the Ukrainian market.

Since 1998, Carlsberg has also owned the Lvivske Brewery in Lviv, which dates back to a monastery brewery started in 1715. To celebrate this fact, the brewery produces a 4 percent ABV Lvivske 1715 Pilsner. Other products include 4.8 percent ABV Lvivske Zhive, 7 percent ABV Lvivske Mitsne (Strong), and 8 percent ABV Lvivske Porter.

Promise is seen for smaller craft brewing operations in Ukraine. The Russian industry journal *Pivnoe Delo* (*Beer Business*) recently summarized the Ukrainian market by writing that "consumption polarization is the main trend of the market—the economy and premium segments grow due to the sales of mass beer sorts. Carlsberg Group and SABMiller seized the opportunity given by this trend and increased their market share yet the mainstream leaders companies AB InBev and Obolon saw some market share decline. . . . In spite of the actions of international companies, the market shares of small producers and import continue to grow."

**Above: A glass of Chernihivske lager on the bar, somewhere in Ukraine.** (Chernihivske)

# The Czech Republic

The Czech Republic came into being in 1993 through the dissolution of Czechoslovakia, but it is comprised of ancient lands with a vibrant beer and brewing culture that goes back to well before the earliest documented breweries that existed here in the tenth century. The Czech Republic consists of the old kingdoms of Bohemia and Moravia as well as a sliver of Silesia. Long ago independent, they had been absorbed into the Austro-Hungarian Empire during the nineteenth century and then became part of independent Czechoslovakia along with Slovakia after World War I, only to find themselves occupied, mauled, and exploited by Germany in the 1940s and by the Soviet Union through the 1980s. During these years, the brewing industry, predominantly located in Bohemia, was destroyed and rebuilt several times, yet it always emerged as one of the most important in continental Europe.

Traditionally, the nexus of Czech brewing was in a trio of Bohemian cities, the capital city, Prague, as well as in Pilsen (Plzen) and Ceske Budejovice.

Of these, Pilsen is the most internationally important, having given its name to the golden pilsners, one of the classic all-barley styles of lager. The pilsner style is well known around the world and is the most emulated style in the world, though it is more often that not badly imitated. The word is usually spelled "pilsner," occa-

Below: A couple in period costume enjoy mugs of pilsner in the fermenting cellars of Pilsner Urquell. The brewery was founded in 1842. (Pilsner Urquell)

sionally spelled "pilsener," and often shortened simply to "pils." While there are many "pilsners," there is only one *original* pilsner. This is the beer, first brewed in 1842, that is known in Czech as Plzenky Prazdroj, but known around the world by the German phrase on the distinctive white and gold label: Pilsner Urquell, which means "Pilsner from the Original or Ancient Source." The explicit Pilsner Urquell trademark was registered in 1898 in a document that referred to "the absurdity and illogic of using the word 'Pilsner' for beer brewed in towns outside of Pilsen." However, the term long ago became generic.

Pilsner Urquell is a 4.4 percent ABV lager made with Saaz hops, a mild and aromatic hop variety which originated in Bohemia and is named for the town of Saaz (Zatec in Czech). Saaz hops are lower in alpha acids and therefore less bitter than the hops favored by American craft brewers. It was one of the first lagers, beers made with bottom-fermenting yeast, and is perhaps the oldest major lager in the world. Pilsner Urquell was first developed by Martin Kopecky, who supported the idea of producing an outstanding bottom-fermented beer that could replace many top-fermented beers in Pilsen pubs. Production started in October 1842, and the first barrels were shipped to Pilsen restaurants. Several were sent to Karel Knobloch's pub on Liliova Street in Prague, which was the first commercial draft account for the new beer. Pilsner Urquell was also a success in Bohemian spa towns such as Karlovy Vary (Karlsbad), and this led to its eventually being distributed worldwide.

By 1856, the beer from Pilsen was being served in Vienna, and three years later it had reached Paris. Beginning in October 1900, the Pilsen "beer train" left early every morning for

Below: Pouring a pilsner at Restaurace Na Spilce, the restaurant in the courtyard of the Pilsner Urquell brewery in Pilsen. (Pilsner Urquell)

Above: The main gate at the Pilsner Urquell brewery. The clock celebrates the 170th anniversary of the brewery. (Pilsner Urquell)

Below: Pilsner Urquell and beers from other SABMiller Czech breweries, Gambrinus, Radegast, and Kozel light and dark. (SABMiller)

Vienna. A similar train also went to Bremen. By 1874, this Czech beer had arrived in America.

The path to the world had begun. As Pilsner Urquell became established in European cities from Lvov to London, it began to receive medals and honors

from international shows and exhibitions.

The brewery at Pilsen reached an annual production of over a million hectoliters before World War I, but that number would not be exceeded for many decades because of the world wars. In the late 1930s, production rebounded, and the Pilsner breweries accounted for 75 to 85 percent of the entire beer export from the country. Production levels fell again during World War II. The Communist government seized the brewery after the war, but by 1956 Pilsner Urquell was once again being exported, bringing in much needed hard currency for the state government.

Successive plant modernizations between 1965 and 1985

prepared the original Pilsner for the inevitable market expansion that followed the collapse of Communism in 1989. This event also opened the door to the potential of a wider range of Czech beers reaching the Western market. The company was privatized, and in 1999 a controlling interest was acquired by South African Breweries (SAB), which became SABMiller in 2002.

The second largest of the great Czech brewing companies, Pivovary Staropramen, located in Prague's Smichov district, is named for an old spring from which it drew its water. Beer was first commercially brewed in Smichov in 1871, but it was the Ostravar and Branik breweries, founded in 1898–1899, who merged and named themselves Staropramen in 1911.

The largest brewery in Prague in the 1930s, Staropramen was nationalized after World War II and privatized in 1989. In 1992, it joined with the Branik and Mestan breweries to form Prazske Pivovary (Prague Breweries Group), which was acquired in 1996 by Bass from the United Kingdom, who also bought Ostravar in 1997. In 2000, Interbrew (AB InBev after 2004) bought Bass and at the same time made extensive investments in Eastern Europe. Most of the latter were divested in 2009 to the London-based private equity fund CVC Capital Partners, who organized their holdings into an entity called StarBev, named after Staropramen, which was their flag-

Above: Discussing the finer nuances of the beer at the annual Pilsner Fest in Pilsen. (Pilsner Urquell)

Below: The brewing hall at Pilsner Urquell. (SABMiller/One Red Eye)

Above: A brewery worker with a glass of Pilsner Urquell poured directly from a keg in the cellars at the Pilsen brewery. (One Red Eye/Philip Meech)

Below: A historic photo of a delivery truck, circa the 1920s. (Pilsner Urquell)

ship holding. CVC sold StarBev to Molson Coors in 2012.

The pilsner-style beers brewed in Prague at Staropramen include Staropramen Lager, Staropramen Lezak, and Vratislav, all at 5 percent ABV. Additional pilsners are 4 percent Staropramen Svetly and 5.3 percent Staropramen Velvet. Among dark lagers are 4.4 percent

Staropramen Dark and 4.8 percent Staropramen Granat. Finally, the brewery produces the Branik brand as a 4.1 percent pilsner.

The third pillar of Czech brewing is the city of Ceske Budejovice, which was known in German as Budweis during the days of the Austro-Hungarian Empire in the late nineteenth century. The beer of Budejovicky Budvar, the city's major brewery was—and still is—known as Budweiser. In 1876, Adolphus Busch of Anheuser-Busch in St. Louis, Missouri, selected the beers of this region as the model for the national brand he introduced in the United States in 1876, which he named Budweiser. Meanwhile, several other breweries in the United States also used the name to identify a style, just as breweries the world over have used the word "pilsner" to identify a style. Early in the twentieth century, the Missouri and Bohemian breweries reached an agreement by which the Bohemian Budweiser would be sold in the United States under the name "Crystal." The Bohemian beer was exported in Europe, especially in Germany and Austria, as Budweiser, but by the mid-twentieth century, Anheuser-Busch was seeking to establish its own brand there. A trademark dispute ensued in the Court of Justice of the European Union in Case T-191/07. This case was resolved in 2009, and the ruling on the subsequent appeal, Case C-214/09 P, came in 2010. The decision was that "Anheuser-Busch may not register the word 'Budweiser' as a [European Economic] Community] trade mark for beer."

Between 2007 and 2012, under an agreement with Budejovicky Budvar, Anheuser-Busch imported and marketed their "Budweiser Budvar" in the United States under the name "Czechvar." After 2012, this contract went to United States Beverage. This beer, under both names, is a 5 percent ABV lager. Another product from Budejovicky Budvar is 4.7 percent ABV Budweiser Budvar Czech Dark Lager.

In addition to the Big Three brands, there are others which the beer traveler is likely to encounter in the Czech Republic. Three that are now brewed under the Pilsner Urquell umbrella are Gambrinus, Kozel, and Radegast. Kozel was first brewed in 1874 in Velké Popovice near Prague by Franz Ringhoffer. Gambrinus, named for the Jan Primus, the traditional Flemish "king of beer," dates to 1869, while Radegast, named for the Slavic deity of hospitality, is brewed at facility built in 1970 by the government in Nosovice in Silesia. All three flagship brands are 5 percent ABV lagers.

Kralovsky Pivovar Krusovice (Royal Brewery of Krusovice) was started in 1581 in the village Krusovice by

## Ceský Pivní Festival

The biggest beer festival in the Czech Republic, the 17-day **Ceský Pivní Festival** (**Czech Beer Festival**) is held annually in May and/or June in Holešovice, north of Prague. Approximately 70 Czech brewers, as well as international companies, participate. Britain's *Financial Times* considers it as one of the 40 "must attend" global events. In Pilsen, the annual **Pilsner Fest** occurs annually in October. In recent years, festivals of Czech beer have been held in Germany and elsewhere in Europe. (Photos from Ceský Pivní Festival, SE)

Above: Pouring a beer in the cellar at Budejovicky Budvar in Ceske Budejovice. (Budvar)

Below: Budvar lagers are marketed under the Budejovicky, Budweiser and Bud names. Tmavý Leák in the center is their 4.7 percent ABV dark lager. Pardál is another Budvar lager brand (Budvar)

Jiri Birka and later brewed for Rudolf II, King of Bohemia and Holy Roman Emperor. In 2009, it merged with Pivovar Starbrno in the Moravian city of Brno. The merged entity became part of Heineken that same year, but both brands

remain and continue to be brewed in the original breweries. Well known in the export market is Krusovice Cerne, a dark lager found at both 3.5 and 3.8 percent ABV. Among the pilsner-style lagers are 4.2 percent Krusovice Poradna Desitka and 4.7 percent Krusovice Jubilejni Lezak. Starbrno offers a 5 percent ABV pilsner as well as a 3.8 percent Starbrno Cerne black lager and a 6 percent märzen called Starobrno Cerveny Drak.

Another Czech consortium is the Liberec Investment Fund in the Liberec region of northern Bohemia. Through 2009, they had acquired the Nachod, Rohozec, and Svijany breweries, located in the Liberec villages of the same

names. The breweries date from 1872, 1850, and 1564, respectively.

The independent Rodinny Pivovar Bernard in Humpolec dates back to the sixteenth century. By 1991, after postwar state ownership, it was a bankrupt shell of its former glory when it was acquired at auction by Stanislav Bernard, Josef Vavra, and Rudolf Smejkal. Its beers have since become well-known throughout the Czech Republic. In 2010, the brewery organized the country's first all-women beer club. Bernard's portfolio includes pilsners such as 4.7 percent ABV Bernard Svetly Lezak and 5 percent Bernard Svatecni Lezak as well as 5.1 percent Bernard Dark black lager and Bernard Jantarovy Lezak, a German-style dunkel.

For the beer traveler, no visit to the Czech Republic is complete without a visit to Pivovar U Fleku in Prague. Founded in 1499, it is probably the world's oldest brewpub. It has long been a mecca for beer connoisseurs worldwide, who make pilgrimages here to drink, dine, and enjoy the music in U Fleku's labyrinthine complex of eight beer halls or in its pleasant beer garden. It was acquired in 1762 by Jakub Flekovsky and became known by its present name, meaning "at Fleks' place."

The only beer available, and a beer available only at U Fleku, is the 5 percent ABV Flekovsky Tmavy Lezak (Flek's Dark Double Lager). It is a pretty good beer, but when enjoyed in the remarkable atmosphere of U Fleku, it becomes exceptional.

**Above: The brewing hall at Budejovicky Budvar. (Budvar)**

**Below: The Rytírský Sál (Knight's Lounge), one of eight halls at U Fleku, can accommodate 150 guests. It was formerly a chapel and a malt house. Names of brewers who worked here between 1499 and 1883 can be found on doors. (U Fleku)**

# Germany

**G**ermany has long been one of the most important brewing nations in the world and the top beer-producing nation in Europe. The German brewing industry was also long noted for its strict adherence to the Reinheitsgebot, the "German Beer Purity Law" or "Bavarian Purity Law." First promulgated in 1516 in the city of Ingolstadt in Bavaria, it was adopted throughout Germany. In fact, Bavaria insisted on its adoption as a precondition for joining the German Empire in 1871. Specifically, the law restricts the ingredients permitted in beer to water, barley, and hops. No mention was made of yeast because the role of that microorganism was neither known nor understood until the work of Louis Pasteur in the nineteenth century. As a means of restricting imports into Germany, the Reinheitsgebot was struck down in a 1988 European Court of Justice ruling, but for beer brewed within Germany, German brewers generally still abide by the Reinheitsgebot as a matter of pride—and because it is good advertising to describe their beers as being brewed under this ancient regulation.

Beer styles in Germany are not as varied as they are in Belgium or the United States. Lager is overwhelmingly dominant here, as in most countries, but German brewers do produce a range of lagers. Most brewing companies offer a range that includes pale lagers (helles), dark lagers (dunkel), seasonal beers,

**Below: Friends enjoy a round of Krombacher pilsners. Ceramic tap towers, such as the one in the background, are a common fixture in German drinking establishments. (Krombacher)**

(including märzen), bocks, and double bocks (doppelbocks).

Wheat beers are increasingly popular and are known interchangeably as weiss (white) or weizen (wheat) bier. Hefeweizen, now copied worldwide, is literally a wheat (weizen) beer that is unfiltered and therefore still has yeast (hefe) in suspension. Wheat beer is brewed throughout Germany with specific styles indigenous to northern Germany, Bavaria, and especially Berlin, where Berliner Weisse is the city's trademark beer.

The product portfolios of German brewing companies also include non-alcoholic (alkoholfrei) lager brands, which, like such products anywhere, can actually have up to 0.5 percent ABV. The alkoholfrei beers are promoted with a higher profile than is usually accorded such products in other countries. As elsewhere in Eastern and Central Europe, many German brewers produce radler, the popular beer and carbonated fruit juice blend.

In 1930, Germany produced 58 million hectoliters from 4,669 plants compared to 20 million hectoliters from 610 plants in Britain and no production in the United States because of Prohibition. World War II crushed all German industrial production, but within five years, 2,637 breweries had reopened and were brewing 10 million hectoliters. By 1965, production was at 66 million hectoliters from 2,125 breweries. Until the early 1970s, most beers sold in Germany were brewed in a local brewery, but thereafter a move was made toward creating more nationwide brands.

Because of consolidation, the number of breweries declined to around 1,300 by 1990 and has fluctuated between 1,250 and 1,300 since. Total output peaked at 120 million hectoliters in 1992 and has declined since, with the German brewing industry producing 96 million hectoliters in 2010. Nearly

**Above: A vintage photo taken at the Warsteiner brewery near the Arnsbergerwald in Westphalia in northwest Germany. (Warsteiner)**

**Below: Enjoying glasses of Diebels Alt in Düsseldorf. Altbier, or "beer in the old style," is Düsseldorf's signature beer style. (Diebels)**

**Above: The Köstritzer brewery in Thuringia. (Bitburger)**

**Below: Beers from the Bitburger portfolio include the company's flagship Premium Pils, as well as Köstritzer Schwarzbier and the products of Licher in Lich and König in Duisburg. (Bitburger)**

half of German beer is produced in two states—North Rhine-Westphalia (Nordrhein-Westfalen) with around 24 percent of the total and Bavaria (Bayern) with about 22 percent.

As of press time, the four largest German brewing companies were Oettinger, with an annual output of about 6 million hectoliters; Krombacher, with about 5.5 million; Bitburger, with about 4 million; and Beck's, with about 3 million. Below this, Warsteiner, Hasseröder, and Veltins were in a near tie in the vicinity of 2.75 million hectoliters. Munich's largest brewing company, Paulaner, was next with about 2.3 million, followed by Radeberger and Erdinger with close to 2 million. Unlike the situation in most

countries globally, the three largest German breweries are not part of multi-national conglomerates.

Between 2001 and 2004, the Belgian conglomerate Interbrew (AB InBev after 2008) made a major move into the German market, acquiring Beck's, Diebels, Gilde, and two of Munich's Big Six, Spaten and Löwenbräu, which had merged under the same holding company in 1997.

Oettinger surpassed Krombacher as Germany's largest brewing company in 2004, at least in part because of its policy of marketing budget brands, dispensing with an expensive advertising budget, and handling its own distribution. Well established, it dates back to 1333 and now operates breweries at its head-

quarters in Oettingen, Bavaria, as well as in Braunschweig, Gotha, and Mönchengladbach. The product line includes 4.7 percent ABV Oettinger Pils, 4.9 percent Oettinger Schwarzbier, and 4.9 percent Oettinger Hefeweissbier.

Krombacher, based near Kreuztal in North Rhine-Westphalia, was founded in 1803 by Johannes Haas to serve the needs of his father's tavern. In 1896, the Haas family sold it to Otto Eberhardt, who sold it to Bernhard Schadeberg in 1922. The brewery remains in the Schadeberg family in the twenty-first century. Krombacher exports its beer throughout Western Europe, but half of its exports go to Thailand. The brewery's flagship beer is 4.8 percent ABV Krombacher Pils, the biggest selling pilsner in Germany. While most German breweries included wheat beer in their portfolios long ago, Krombacher Weizen, a 5.3 percent wheat beer, was only added to the line in 2007. Also in the line are 4.8 percent Krombacher Dunkel and 5.5 percent Krombacher Weizen Dunkel.

A traveler to the valleys of Germany's legendary Rhine and Moselle rivers will quickly notice that the distinctive phrase "Bitte ein Bit" ("Please, a Bit") on signs in the towns and cities of this region from Trier to Bonn to Frankfurt. This has been the slogan of Bitburger Bräuerei

Theobold Simon since 1951. The brewery dates its lineage back to the small rural brewery founded in 1817 by Johann Wallenborn near Bitburg in what is now the state of Rhineland-Palatinate. In 1842, after Wallenborn's death, Ludwig Simon married into the family and took over the brewery, which was renamed Simonbräu. The present name dates to 1876, when Ludwig's son, Theobald, inherited the company and took the reins of the business. In 1883, the brewery had the distinction of being the first brewer outside Bohemia to brew pilsner. This is now the flagship 4.8 percent ABV Bitburger Premium Pils. Their 0.5 percent alkoholfrei Bitburger

**Above:** The products seen in this Krombacher family portrait are typical of the stylistic lineup at many German breweries, which market a range that includes a dunkel (dark), a pilsner, a weizen (wheat beer), an "alkoholfrei" selection (in this case a weizen), and a radler, the popular blend of lager and lemonade. (Krombacher)

**Below:** A brewer tends his brew kettle at the Krombacher Brauerei in Kreuztal. (Krombacher)

Above: On the bottle line at Beck's in Bremen. (Beck's)

Below: Beers from the German holdings of AB InBev include Beck's, Hasseröder of Wernigerode, and Diebels Alt from Düsseldorf. (AB InBev)

Right: The .33-liter bottle and .5-liter "Steinie-Flasche" used for Veltins Pils. (Veltins)

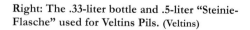

Drive is widely marketed and exported, notably to the United States.

Köstritzer, located in Bad Köstritz in the state of Thuringia, dates to 1543 and found itself in the Soviet occupation zone (later East Germany) after World War II. The brewery was acquired by Bitburger in 1991 after German reunification, and now contributes about a quarter of Bitburger's annual output. The brewery is famous for its 4.4 percent ABV Köstritzer Schwarzbier. Indeed, the brewery helped to originate and popularize the schwarzbier (black lager) style and is now the market leader in Germany in this category. In the nineteenth century, the author and poet Johann Wolfgang von Goethe was a fan of Köstritzer Schwarzbier, and famously drank it for a time when he was too sick to eat solid food. In an 1892 letter, Prince Otto von Bismarck wrote that Köstritzer holds "a distinguished rank within the aristocracy of beers."

Bräuerei Beck, known simply as Beck's, is Germany's fourth-largest brewer and has long been the largest exporter of German beer, in part because of its location in the city-state of Bremen, which has been an important seaport since the days of the Hanseatic League more than 600 years ago. The company was started in 1873 by Heinrich Beck, Thomas May, and Lüder Rutenberg, and was known as Kaiserbräuerei Beck & May until 1875, when May left the business and it was renamed as Kaiserbräuerei Beck & Co. Long a family-owned company, Beck's was acquired in 2002 by InBev (AB InBev since 2008). Beginning in 2012, the Beck's products sold in the United States have been brewed by AB InBev in St. Louis, Missouri.

The flagship beer, brewed in Bremen and marketed simply as Beck's, is a 5 percent ABV pilsner. There are also a number of other products, all lagers, including 4.8 percent Beck's Dark, 4.9 Beck's Gold, 2.3 percent

Beck's Light, and 5 percent Beck's Oktoberfest.

Also brewed by Beck's in Bremen is the brand St. Pauli Girl, which originated at the St. Pauli Bräuerei, so named because of its original location near the former St. Paul's Monastery. The brewery was founded in 1857 by Lüder Rutenberg, later one of the original founders of Beck's (as noted above). First sold in the United States in 1965, St. Pauli Girl is a 4.9 percent pilsner produced only for export sale and is not sold in Germany. Its labels are known for depicting a buxom woman carrying beer steins. Beginning in 1982, the labels began carrying a picture of an actual woman, whose image was used for one year. The "St. Pauli Girls" have alternated between actresses, fashion models, and Playboy Playmates.

Warsteiner was founded in 1753 by Antonius Cramer as a farmhouse brewery, and though it has grown into one of Germany's largest companies, it is still owned by the Cramer family and is Germany's largest family-owned brewing company. In 1803, Caspar Cramer built the original building, the Warsteiner Domschánke, a small brewpub in the centre of the village of Warstein in what is now North-Rhine Westphalia. The building still serves as the company headquarters, while the brewery is located in the picturesque Arnsberg Forest Nature Park

## German Beer Festivals

**Oktoberfest** in Munich is the world's largest beer festival, attracting around six million attendees annually. Oktoberfest originated as a celebration of the wedding of Bavaria's Crown Prince Ludwig (later King Ludwig I) and Princess Therese of Saxe-Hildburghausen on October 12, 1810. The festivities took place in Munich in the field known as Theresienwiese (Theresa's meadow), and are still held there today. Oktoberfest lasts for 16 days, ending with the first Sunday in October, unless that day falls before German Unity Day (October 3),

**Above: An aerial view of the Oktoberfest grounds by night. (Paulaner)**

in which case, Oktoberfest goes on for a seventeenth day. This original Oktoberfest has inspired countless other Oktoberfests across Germany and throughout the world.

The **Cannstatter Volksfest**, held at Cannstatter Wasen in Stuttgart, is the second largest beer festival in Germany, hosting four million people annually. Dating to 1818, it lasts for three weeks, ending on the second Sunday in October.

The **Gäubodenvolksfest** in Straubing, attracts more than a million festival-goers and is held for 11 days during the middle of August, when it is considered locally to be the "fifth season." The **Barthelmarkt** in Manching, near Ingolstadt, is held during the last weekend of August, and dates back to at least 1354. The **Bergkirchweih** in Erlangen has taken place since 1755, beginning on the Thursday before Pentecost (seven weeks after Easter). A 12-day festival, it attracts about a million visitors and features the largest open-air biergärten in Europe, with seating for 11,000.

The **Internationales Berliner Bierfestival** (**International Berlin Beer Festival**), which was started in 1997, is a three-day event taking place in early August. It features the popular "Berlin Beer Mile," running from Strausberger Platz to the Frankfurter Tor in the center of Berlin, which is lined with the stands of more than 300 breweries from about 100 countries.

Above: An aerial view of Brauerei C & A Veltins in Meschede-Grevenstein. (Veltins)

Below: The Radeberger Group portfolio includes the flagship pilsner, as well as the beers of the Dortmunder Actien Bräuerei (DAB) and Henninger of Frankfurt, both one-time major independent brewers. Widely exported Clausthaler is the world's leading alcohol-free beer. (Radeberger)

outside of Warstein. As the company notes, "In 1927, the discovery of the Kaiserquelle — a natural reserve of extra-soft water in the Arnsberg forest—signaled a change in the production process. The soft water of the Kaiserquelle today still feeds the water tanks of the Warsteiner Brewery, and contributes to the unique taste experience of Warsteiner beer." The flagship beer is a 4.8 percent ABV pilsner called Warsteiner Premium Verum. Other products include 4.9 percent Warsteiner Premium Dunkel and 2.4 percent Premium Light.

Hasseröder is located in Wernigerode in the state of Saxony-Anhalt. It originated in 1872 as the Zum Auerhahn (To the Grouse) Bräuerei and was owned by Ernst Schreyer from 1882 until it went public in 1896. After World War II, it was confiscated by the East German government. After German reunification in 1991, it went public and was acquired by Interbrew (later AB InBev). Products include 4.9 percent ABV Hasseröder Premium Pils, 5 percent Hasseröder Schwarz, and 5.8 percent Hasseröder Fürstenbräu Granat.

Bräuerei C&A Veltins is located in Meschede-Grevenstein in the state of North Rhine-Westphalia. It originated in 1824 with Franz Kramer as a brewpub

and evolved as a larger brewery that was acquired in 1852 by Clemens Veltins. The initials used in the name are those of the twins, Carl and Anton Veltins, who took over from their father in 1893. The company is still owned and managed by the Veltins family, who brew their 4.8 percent ABV Veltins Pilsner.

Radeberger originated in 1872 as the Bergkeller in Radeberg, near Dresden, in what is now the German state of Saxony. It has a long history of being a favorite of political leaders. In 1887, Chancellor Otto von Bismarck declared Radeberger's pilsner as the "Kanzler-Bräu" (Chancellor Beer), and it was also a favorite of Frederick Augustus III of Saxony. After World War II, Saxony fell into East Germany and the brewery fell under state control, but it still attracted an elite clientele. According to Peter Truscott, a biographer of Vladimir Putin, the future president of Russia drank Radeberger when he was a KGB agent in Dresden in the 1980s.

After German reunification, Radeberger was acquired by the Binding Bräuerei. The largest brewery in Frankfurt, located in Germany's financial capital Binding, is known for its Clausthaler family of 0.5 percent ABV alkoholfrei beers. Binding was acquired in 1994 by the Oetker Group, a diversified holding company that owns a wide range of interests from hotels to

insurance to cooking products. The brewery holdings of the Oetker Group are now consolidated within the Radeberger Group and include Binding as well as Brinkhoff in Dortmund, Dortmunder Actien Bräuerei (DAB) in Dortmund, Freiberger in Freiberg, Schultheiss in Berlin, Leipziger Bräuhaus zu Reudnitz (Sternburg) in Leipzig, and Tucher in Fürth.

Products from the extensive portfolio of the Radeberger Group include 4.8 percent ABV Radeberger Pilsner, 4.5 percent Binding Lager, 4.9 percent Binding Römer Pils, 5 percent Kutscher Alt, 5 percent Schöfferhofer Hefeweizen, 4.8 percent DAB Pilsener, 5 percent DAB Original, 4.9 percent Freiberger Premium-Pils, 4.8 percent Dortmunder Kronen Pilsener, 5.8 percent Freibergisch Jubiläums Festbier, 5 percent Schultheiss Pilsner, 6.9

**Above and below: Serving kölsch, the signature beer of the city of Köln (Cologne), at Gaffel am Dom, the Gaffel Becker brewery restaurant in the city center. Gaffel is one of the important practitioners of the kölsch-brewer's art, and their restaurant is described as a place "where friendly people meet."(Gaffel)**

Above: Two glasses of Ganter pilsner, ready to serve. (Ganter)

Below: Pilsner and Weizen from Brauerei Paderborner. The brewery was founded in Paderborn in 1852, and acquired by Warsteiner in 1990. (Paderborner)

percent Reudnitzer Heller Ur-Bock, 4.9 percent Sternburg Pilsener, 4.9 percent Sternburg Schwarzbier, 5 percent Tucher Pilsener, 6 percent Tucher Bergkirchweih Festbier, 5.1 percent Tucher Kristall Weizen, and 7.4 percent Tucher Bajuvator Doppelbock.

Representing the group's solid footing in Dortmund, the line includes such export style lagers as 5.3 percent ABV Binding Export, 5.4 percent DAB Export, 5.1 percent Dortmunder Kronen Export, 5 percent Hansa Export, 5.2 percent Sternburg Export, and 5.3 percent Dortmunder Union Export.

It should be mentioned at this juncture that "export" is actually a lager style indigenous to Dortmund. They are traditionally full-bodied but not quite as sweet as the beers of Munich or as dry as a true pilsner. This style evolved when they began transporting beer to other markets across the continent. In order to withstand the rigors of travel, Dortmund brewers produced a beer that was well hopped and slightly higher in alcohol. As such, the Dortmund lagers came to be known by the name "export."

Among Germany's largest brewing companies is the Dortmund-based holding company Bräu und Brunnen AG, which was the largest in Germany in the 1970s and 1980s, with an annual output of up to 15 million hec-

toliters of beer until the 1990s, when it began slipping into the red and shifting its focus more to soft drinks. It was created in 1972 by the merger of Schultheiss Bräuerei, the largest brewery in Berlin, and the Dortmunder Union Bräuerei (DUB). The latter was formed in 1873 by the merger of several small breweries and was ranked among the top 10 German breweries for half a century before the merger. Other Bräu und Brunnen holdings included Dortmunder Ritter in Dortmund; Heidelberger Schlössguell Bräuerei (home of the famous Valentins Weizen Beer) in Heidelberg; and Einbecker Bräu Haus in Einbeck, the oldest bock beer brewery in Germany. Since 2004, the Bräu and Brunnen brewery holdings were spun off into the Oetker Group.

As export is the beer of Dortmund, other beer styles which are identified with specific German cities are altbier ("old style beer") in Düsseldorf and kölsch in Cologne (Köln). Roughly the German equivalent of English or American ale, altbier is literally a beer made in the "old" way (pre-nineteenth century) with top-fermenting yeast. Typical of alt brewers is the Gatzweiler family, who have been brewing altbier since 1313 and operated the Zum Schlussel Haus Bräuerei (a brewpub) in Düsseldorf from 1937 to 1944, when it was destroyed in World

War II. Zum Schlussel was rebuilt after the war and was augmented by the construction of a large, modern brewery in 1963. Another important brewer of altbier is Bräuerei Diebels, founded in 1878 by the Krefeld brewmaster Josef Diebels in Issum. Beginning in the 1970s, Diebels became the first to take altbier nationwide. In 2001, it was acquired by Interbrew. In 2005, for the first time in decades, Diebels added pilsner to its list. The flagship Diebels Alt is rated at 4.8 percent ABV.

Kölsch, top-fermented with the golden color of a pilsner, has been a "protected designation of origin," since 1997, limiting Kölsch brewing to only Cologne and its environs. In Cologne, two examples of independent kölsch breweries are the Cölner Hofbräu Früh, started in 1904 by Peter Joseph Früh, and Gaffel Becker & Company, founded in 1908 by the Becker

Brothers. Both of them operate inviting brewery restaurants, which are a welcome stop for the beer traveler, in the center of Cologne. They both brew and serve a 4.8 percent ABV kölsch.

Two important brewing centers in the north of Germany are the city-state of Hamburg and the city of Hannover in Lower Saxony (Niedersachsen). Brewer Cord Broyhan made his first beer in

Above: Gaffel am Dom, the Gaffel Becker brewpub on the northern side of the cathedral in Köln (Cologne). (Gaffel)

Below: Brauerei Dinkelacker, the largest brewery in Stuttgart, The brewery Dinkelacker brewery was founded by Carl Dinkelacker in 1888. Their Privat is an 5.1 percent ABV export-style lager. (Dinkelacker)

Above: Toasting Ganter's new brewhouse are Manager Director Hartmut Martin, Otto Steiner of the Steiner Sarnen Agency, Katharina Ganter Fraschetti, director of the brewery, Freiburg Mayor Ulrich von Kirchbach and Ganter CEO Detlef Frankenberger. (Ganter)

Below: Ganter's Pilsner, Export, and three weizens: Hefedunkel, Hefehell, and Alkoholfrei. (Ganter)

Hannover in 1526. Two decades later, the Hannover Brewers Guild (Gilde) was founded, taking as its symbol a seal with Broyhan's initial. The Guild's brewery evolved over the centuries, and in 1968 it took over the 114-year-old Lindener Aktien Bräuerei, which markets its products under the name Lindener Gilde. In 2003, Gilde was acquired by Interbrew (later InBev). The flagship beer is 4.9 percent ABV Gilde Pilsener.

Another important brewer in northern Germany—and even more important in the export market—is the Holsten Bräuerei, which was founded in 1879 at Altona (now part of Hamburg). The Holsten Group, which developed from the Holsten Bräuerei, became one of the biggest brewery groups in Germany, with breweries in Hamburg and Kiel, the Feldschlösschen Brewery in Brauschweig, the Kronen Brewery in Luneberg, the Mecklenburg Brewery in Lubz and the Sachsische Bräu Union (SBU) in Dresden in the former East Germany. Meanwhile, Holsten

beer was produced under license in breweries in Great Britain, China, Nigeria, Hungary, and, since 1992, Namibia. Holsten was acquired by the Carlsberg Group in 2004. Holsten products include 5.2 percent ABV Holsten Pils, 5.2 percent Holsten Export, 7 percent Holsten Festbock, and 7 percent Holsten Maibock.

Dinkelacker-Schwabenbraeu in Stuttgart is the largest brewery in the state of Baden-Württemberg. The brand originated in 1888 when Carl Dinkelacker took over a brewery that dated back to 1824. The product line includes Cluss, a Keller-style lager, and Leicht, a low-calorie beer introduced in 1990. Stuttgart's Schwaben Bräu Robert Leicht AG evolved from the first beer brewed in 1878 by Robert Leicht at his gasthof called Ochsen. The extensive product line includes 5.1 percent ABV Dinkelacker Goldhalsle, 5.6 percent Dinkelacker Märzen, 4.9 percent Schwaben Pilsner, and 5.5 percent Schwaben Weihnacts Bier.

Bräuerei Ganter is a privately owned independent brewery located in Freiburg, in the state of Baden-Württemberg. It was founded in 1865 by Louis Ganter on the site of the 1854 Ringwald Brewery and is now operated by the Erlmeier family. Products include 4.8 percent ABV Freiburger Pilsner, 4.9 percent Ganter Pilsner, and 7.5 percent Ganter Wodan, a doppelbock.

# BAVARIA

When it comes to beer, the state of Bavaria is to Germany what Germany is to the rest of the world. Over half of Germany's breweries are in Germany's largest state, and Munich (München), the state capital, is one of the cities that is most celebrated in global beer culture. Munich's beer festival, Oktoberfest, has been held annually (except during the world wars) since 1810. Lasting nearly three weeks, Oktoberfest attracts around six million attendees each year, including beer travelers from around the world, and is emulated in countless Oktoberfests occurring across the globe.

It was also in Munich that Gabriel Sedlmayr at the Spaten Brewery first exploited lager yeast (*Saccharomyces pastorianus*, formerly *Saccharomyces carlsbergensis*) in the early nineteenth century. This development, together with practical refrigeration, revolutionized the history of brewing and gave lager to the world.

In Munich, there are an institutionalized Big Six brewing companies which survived until late in the twentieth century as independent breweries. They are Augustiner, Hacker-Pschorr, Hofbräuhaus, Löwenbräu, Paulaner, and Spaten-

**Above: A young "Paulina" in a Bavarian-style dirndl dress serves up Paulaner Oktoberfest Märzen at the Paulaner Biergärten on the Nockherberg in Munich. In most Munich biergärtens, the one-liter stein is the default glassware, although weissbier is served in distinctive weissbier glasses such as the one seen atop page 97.**

**As the company's literature reflects, the Paulaner Biergärten "has long been an institution and a part of Munich's cultural history . . . more than just a place: it is an attitude to life. It represents indulgence and conviviality, tradition and friendship." The same, naturally, can be said of the biergärtens and beer halls of the rest of Munich's Big Six. (Paulaner)**

Above: Enjoying a beer on a warm summer day. (Paulaner)

Below: Erdinger Weissbräu is Germany's largest producer of wheat beers. The portfolio includes Erdinger Weissbier, Erdinger Dunkel, Erdinger Weissbier Pikantus, Erdinger Schneeweisse, and Erdinger Kristall, a 5.3 percent ABV kristalweizen, or filtered wheat beer. (Erdinger)

Franziskaner. By Munich ordnance, only these six breweries within the city limits are invited to serve their beer at Oktoberfest.

An aspect of Munich beer culture that is of interest to the beer traveler, whether on site in Munich or at a specialty beer shop at home, is the tradition of assigning names ending in "ator" to high-alcohol doppelbocks. These include the 7.5 percent ABV Hofbräu Delicator, the 7.6 percent Löwenbräu Triumphator, the 7.6 percent Spaten Optimator, the 7.9 percent Paulaner Salvator, and the 8.1 percent Hacker-Pschorr Animator. From just outside Munich comes the 6.7 percent Ayinger Celebrator. A bit farther afield, and an eisbock not a doppelbock, Schneider's Aventinus Weizen-Eisbock is

worth mentioning because it is the most extreme of the lot, weighing in at 12 percent ABV.

Augustiner was founded in 1328 by monks of the Order of St. Augustine and acquired in 1829 by Anton and Therese Wagner, whose family still owns a controlling share, making it Munich's oldest independent brewery. It is said to be the most popular Munich beer in Munich. The concise lineup includes 4.6 percent ABV Augustiner Bräu Märzen Bier and 6.5 percent Augustiner Bräu Christmas Bock Bier.

Hacker-Pschorr is a brewery with two names that have been intertwined since brewmaster Joseph Pschorr married into the Hacker family in 1793. The Hackers had been brew-

ing since 1414, hence the date on the Hacker logo. In 1834, Joseph Pschorr bequeathed his brewing empire to his two sons, with Georg Pschorr heading Pschorrbräu and Matthias Pschorr taking Hackerbräu. They were cross-town rivals until 1944, when the Pschorr operation was put out of commission in an air raid, and Hacker came to their aid. The two firms officially became Hacker-Pschorr Bräu in 1972, which became part of the Paulaner Group in 1985.

The portfolio includes 5.5 percent ABV Hacker-Pschorr Münchner Gold, 5 percent Hacker-Pschorr Münchner Dunkel, 5.5 percent Hacker-Pschorr Hefe Weisse Naturtrüb, 5.3 percent Hacker-Pschorr Dunkel Weisse, and 5.8 percent

Hacker-Pschorr Oktoberfest-Märzen.

Hofbräuhaus (HB) is a brewing company whose Munich beer hall of the same name is one of the largest and most famous drinking establishments in the world. HB was started in 1589 as the household brewery for Wilhelm V, the Duke of Bavaria.

All of the Big Six breweries have vast beer halls, but HB has an ethos that stands it apart. The HB lagers are brewed off site, but the beer hall remains as perhaps the top attraction for beer travelers in the city that likes to think of itself as the world capital of beer drinking. The traveler visiting HB will have a choice that includes 5.1 percent ABV Hofbräu

Above: Coasters representing Munich's traditional Big Six: Augustiner, Hacker-Pschorr, Hofbräuhaus, Löwenbräu, Paulaner, and Spaten. These six are followed by an early solo Hacker coaster, and one representing Salvator, Paulaner's doppelbock. (Author's collection)

Below: Friends, decked out for Oktoberfest, enjoy some camaraderie at a biergärten. (Paulaner)

**Above: Servers with six or eight steins are a common sight at Munich beer halls. (Paulaner)**

**Below: The Paulaner portfolio includes Oktoberfest Märzen, Original Munich, and Pilsner. Next come Paulaner's Thomasbräu Alkoholfrei and Salvator doppelbock, lowest and highest in alcohol within the portfolio. (Paulaner)**

Original Helles, 5.5 percent Hofbräu Dunkel, 5.1 percent Hofbräu Münchner Weisse, and 6.3 percent Hofbräu Oktoberfestbier.

Löwenbräu (Lion's Brew) has been a Munich institution since around 1383. Some legends trace the name to a brewpub called Zum Löwen, and others refer to a painting of Daniel in the lion's den that was on the brewhouse wall. In any case, Löwenbräu was built into an industrial-scale brewery between 1826 and 1855. One of Munich's largest, it led the way in exports and in overseas licensing, including a deal in the 1970s whereby it was brewed

in the United States by Miller. For the beer traveler visiting Munich, the lineup includes 5.2 percent ABV Löwenbräu Original Helles, 5.4 percent Löwenbräu Urtyp Helles, 5.5 percent Löwenbräu Premium Dark, 5.2 percent Löwenbräu Schwarze Weisse, and 6 percent Löwenbräu Oktoberfestbier.

Paulaner dates back to the first beer brewed in 1631 by the monks of the Order of St. Francesco di Paola. The monks sold their beer to the public after 1780, but the brewery became a secular lager brewery early in the nineteenth century. In 1928, the brewery merged with the Gebrüder Thomas (Thomas Brothers) brewery creating Paulaner-Salvator-Thomasbräu, which was officially simplified to "Paulanerbräu" in

1994. Salvator is Paulaner's doppelbock, created by brewer and owner Franz Xavier Zacherl in 1861. In 1985, Paulaner acquired Hacker-Pschorr, and it added Kulmbacher in 1994. In 2011, Paulaner relocated from its Nockherberg brewery in Munich to the Munich suburb of Langwied.

The Paulaner portfolio includes 4.9 percent Paulaner Original Münchner, 5 percent Paulaner Original Münchner Dunkel, 5.5 percent Paulaner Hefe-Weissbier Naturtrüb, and 5.8 percent Paulaner Oktoberfest-Märzen. The Thomasbräu name is recalled in 6.8 percent Paulaner St. Thomas Bock and 0.5 percent Thomasbräu, an alkoholfrei selection.

Now part of Paulaner, the Erste Kulmbacher Union (EKU) was an amalgam of smaller breweries in the Kulmbach area that joined together in 1872 to brew Echt Kulmbacher Pils, which is still the EKU flagship beer. Kulmbach is a town nestled in the rolling foothills of the Bavarian Alps in a picturesque town perched in a valley overlooked by a castle. In 1987, EKU had absorbed Henninger, Frankfurt's big brewing company. Begun in 1869 by Christian Henninger, it developed a strong market presence in the Mediterranean area. Today, the Kulmbacher family includes 4.9 percent ABV EKU Hell, 5 percent EKU Pils, 5.9 percent Kulmbacher Lager Hell, 5.6 percent Kulmbacher

Festbier, 9.2 percent Kulmbacher Reichelbräu Eisbock, 5.4 percent Kulmbacher Export, 5.4 percent Kapuziner Weissbier, 4.9 percent Mönchshof Bayerisch Hell, 4.9 percent Mönchshof Schwarzbier, 5.6 percent Mönchshof Weihnachts Bier, and 6.9 percent Mönchshof Maibock.

The Spaten portion of Spaten-Franziskaner-Bräu traces its roots to 1397, but the initials on the logo are those of Gabriel Sedlmayr, master brewer to the royal court of Bavaria, who bought the brewery in 1807, and of his son, also Gabriel Sedlmayr, who essentially invented lager. The largest brewery in Munich during the last half of the nineteenth century, Spaten acquired Franziskaner in 1922.

The Spaten lineup includes 5 percent ABV Spaten Pils, 5.2 percent Spaten Münchner Hell, 5.5 percent Spaten Dunkel, and 5.9 percent Spaten Oktoberfestbier Ur-Märzen.

Above: Franziskaner Weissbier, served in a traditional weissbier glass. (Spaten-Franziskaner)

Below: Munich's landmark Hofbräuhaus. (Hofbräuhaus)

distinct brand identities. The flagship brand, 5 percent ABV Franziskaner Weissbier, holds the global number three position in the Weissbier market. Other products include Franziskaner Club-Weisse Kristalklar and Franziskaner Hefe-Weisse Dunkel, both rated at 5 percent ABV.

Today, the Big Six are such in name only, having been reduced though mergers and acquisitions to just four. In 1985, Paulaner absorbed Hacker-Pschorr, though kept its brand name. In 2002, Paulaner became part of Bräu Holding International, a joint venture of Schörghuber Ventures and Heineken. In 1997, Spaten merged with Löwenbräu, to form the Spaten-Löwenbräu-Gruppe, which was acquired by Interbrew in 2003. A year later, Interbrew

**Above: Two old friends at the Aying biergärten. (Bräuerei Aying)**

**Below: The Aying portfolio includes Altbairisch Dunkel, Ayinger Bräu Weisse, Ayinger Oktoberfest-Märzen, Ayinger Ur-Weisse, and Celebrator doppelbock, with its little white goat neck hanger. (Bräuerei Aying)**

Franziskaner is referenced in documents dating back to 1363 as the "brewery next to the Franciscan monastery," and it was owned by Seidel Vaterstetter. In 1861, it was acquired by Joseph Sedlmayr, the son of Spaten's Gabriel Sedlmayr. In 1922, the two breweries merged but retained

merged with AmBev to form InBev, which is now AB InBev.

The third largest brewery in Bavaria after Krombacher and Paulaner is Erdinger Weissbräu in Erding. As the name suggests, Erdinger brews weissbier and is the world's largest producer. The brewery was founded in 1886 by Johann Kienle, and after several changes in ownership, the general manager at the time, Franz Brombach, bought the brewery in 1935 and renamed it Erdinger Weissbräu in 1949. In 1965, the son of Franz Brombach, Werner Brombach, entered the business. He took over the brewery, which is still family owned. The Erdinger wheat beer family includes 5.6 percent ABV Erdinger Weissbier, 5.6 percent Erdinger Schneeweisse, 5.3 percent Erdinger Weissbier Dunkel, 5.7 percent Erdinger Oktoberfest Weissbier, and 7.3 percent Erdinger Weissbier Pikantus.

Just outside Munich, the Privatbräuerei Franz Inselkammer in village of Aying traces its roots to 1878, when Johann Liebhard started a small tavern brewery. In 1910, the business passed to Liebhard's daughter, Maria, and her husband August Zehentmair. In the 1920s, despite the economic crisis then ongoing in Germany, they built their brewpub, Bräuereigasthof Aying, and began exporting their beer to Munich, nearly 20 miles away.

When August Zehentmair died in 1936, his eldest daughter, Maria, and her husband, Franz Inselkammer, took the reins. After the trying times of the war years, business rebounded to the point where Franz was able to buy the Platzl Hotel in Munich, across the street from the Hofbräuhaus, and turn it into an outlet for his beer. Their grandson, also called Franz, is the seventh generation of his family to preside over the kettles in Aying.

The portfolio includes 4.9 percent ABV Ayinger Bräu Hell, 5 percent Ayinger Altbairisch Dunkel, 5.1 percent Ayinger Bräu Weisse, 5.8 percent Ayinger Oktoberfest-Märzen, 5.8 percent Ayinger Ur-Weisse, 6.7 percent Ayinger Winterbock, 7.1 percent Ayinger Weizenbock, and a 5.8 percent Oktober Fest-Märzen.

**Above: Guest at a Paulaner tent at Munich's Oktoberfest. The Winzerer Fändl, the largest of Paulaner's seven tents, seats 8,450 and is Oktoberfest's biggest. (Paulaner)**

**Below: The Hofbräuhaus biergärten is located adjacent to their great Munich beer hall. (Hofbräuhaus)**

**Above: Georg Schneider VI is the seventh generation proprietor of G. Schneider & Sohn in Kellheim, north of Munich.** (Schneider)

**Below: The lineup includes Schneider Weisse, Kristall, Mein Grünes, Unser Aventinus, and Aventinus Eisbock.** (Schneider)

I and his son Georg Schneider II took over the Weisses Bräuhaus in Munich, then the oldest wheat beer brewery in the city. In 1927, Georg Schneider IV opened a second brewing facility in Kelheim, about 60 miles north of Munich. After the Munich brewery was damaged during World War II, brewing operations were consolidated in Kelheim, where they remain, now under the hand of Georg Schneider VI. The brewery continues to operate two restaurants, the Weisses Bräuhaus in the center of Munich as well as another in Kelheim.

Another seventh-generation family-owned-and-operated brewery near Munich is G. Schneider & Sohn (aka Schneider Weisse), which specializes in light and dark wheat beers. The business dates to 1872, when Georg Schneider

Schneider hefeweizens include 3.3 percent ABV Schneider Weisse Tap 11 Unsere Leichte Weisse, 4.9 percent Schneider Weisse Tap 1 Mein Blondes, 5.5 percent Tap X Meine Sommer

Weisse, and 6.2 percent Schneider Weisse Tap 4 Mein Grünes. Weizenbocks include 7.3 percent Schneider Weisse, 8.2 percent Schneider Weisse Tap 6 Unser Aventinus, and 12 percent Aventinus Eisbock.

Munich may be the capital of Bavarian brewing, but it is not the only city in the state with a unique beer culture. Bamberg, about 125 miles to the north of Munich, is the traditional home of rauchbier, which is literally "smoked beer." The malted grain used in the brewing of rauchbier is roasted over the open flame of a beechwood fire. The grain is not only roasted to a dark color, but it takes on a distinctive smoky flavor as well. From the sixteenth century until 1991, when the Alaskan Brewing Company took a gold medal at the Great American Beer Festival with their Smoked Porter, smoked beers were essentially unknown outside Bamberg. Today they are still rare, although they are emulated in small batches by brewers in many countries. Smoked beer is generally served with meals including smoked or barbecued meats, rye bread, and certain sharp cheeses. Established in Bamberg in 1718, the Kaiserdom Bräuerei is world famous for its definitive rauchbier.

Another unique and extraordinary beer from Bavaria is steinbier (stone-brewed beer), also spelled "steinbiere." For many centuries, heated stones were the only method of heating large quantities of liquid.

Stone-brewed beers were most often produced in Alpine regions, including Bavaria, Scandinavia, and the mountainous Carinthian region of Austria, where the special stones used for the brewing process could be easily quarried. For brewers, this method made it possible to brew a top-fermented beer with a unique flavor.

When large breweries were built in the nineteenth century, the tedious process of making stone-brewed beers became completely forgotten until 1982, when Privatbräuerei Franz Joseph Sailer, in Marktoberdorf, Bavaria, began brewing the Rauchenfels Steinbier. In this process, stones are heated over a beechwood fire and then dropped into the brew kettles. The result is a beer which, like rauchbier, has a unique, smoky flavor. Bräuerei Hofstetten in St. Martin, Austria, brews a 7.3 percent ABV steinbier called Hofstettner Granitbock.

Above and below: The Weisses Bräuhaus on the Tal, Schneider's Munich restaurant, is on the site where founder Georg Schneider I brewed his first Schneider Weisse Original in 1872. (Schneider)

# Austria

As in Germany, with whom Austria shares a language and many aspects of brewing tradition, lager developed a strong following in Austria in the mid-nineteenth century and now dominates the market, although wheat beer, especially hefeweizen, is popular in mountainous western Austria. As with many other breweries in Central and Eastern Europe, the portfolios of Austria's brewing companies often include radler, a popular blend of lager and carbonated fruit drinks.

Brewing began in what is now Austria well over a thousand years ago, and was well established by the seventeenth century. The definitive beer of the region, the Vienna style of lager, was first brewed by Anton Dreher at Schwechat near Vienna in the 1840s. As with Germany, but unlike places such as the Netherlands and Denmark, Austria has no dominant national brand but rather several well-established regional brands, which were gradually consolidated into a single entity known as Bräu Union (or Bräuunion"). Headquartered in the city of Linz, this entity now controls most of the brewing in the country.

The oldest component of this holding company is the Hofbräu Kaltenhausen in Hallein-Kaltenhausen near Salzburg, which was started by Johann Else Haimer in 1475. The largest brewery in the Salzburg area in the late nineteenth century, it had declined considerably by 1921. In recent years, it was closed, remodeled, and reopened as "a specialty production centre, in which new beer recipes are

**Below: Raising their glasses in the Bräustüberl restaurant at the Stieglbräuerei in Salzburg, which is "dedicated to the top priority of using fresh, regional products." Also fresh, of course, is the beer. (Stiegl)**

constantly developed. Secondly, it provides a forum for beer culture that is popular amongst both professionals from the food industry and beer fans. And thirdly, it has a highly traditional 'Bräugasthof' (brewery inn), where one can enjoy what one has just seen, while sitting next to the brewing copper."

Some of the wheat beers on tap at the Hofbräu Kaltenhausen include Edelweiss Hefetrub and Edelweiss Dunkel Weissbier, both at 5.5 percent ABV, as well as the 5 percent Edelweiss Weissbier Snowfresh spiced beer.

Bräu Union dates back to Bräubank, formed in 1921 through the consolidation of several regional breweries. In addition to Bräuerei Kaltenhausen, these included Linzer Aktienbräuerei, Poschacher, Salzkammergut-bräuerei, and Wieselburger. In 1925, the name was changed to Österreichische Bräu, and through 1929 it absorbed more brewing companies, including Bräuerei Liesing (1928), Aktiengesellschaft Sternbräu Salzburg (1929),

Aktiengesellschaft der Brunner Bräuerei, Brunn am Gebirge (1929), Bürgerliches Bräuhaus Innsbruck (1929), Bürgerliches Bräuhaus Innsbruck (1929), Bräuerei Reutte (1929), and Tiroler Bräuereien Kundl-Jenbach (1929).

In 1939–1945, when Austria was part of Germany, the company was called Ostmarkische Bräu Linz, becoming Österreichische Bräu AG. In 1970, Bräu AG merged with the Bräuerei Zipf in Neukirchen an der Vöckla, which dates back to 1540.

As Bräu AG (later Bräu Union) was taking shape in northern Austria, another regional Austrian brewing group was forming farther south that would be Bräu AG's biggest rival for several decades. Known as Steirische

**Above: The brewmaster at work in the Hofbräu Kaltenhausen microbrewery. (Bräu Union)**

**Below: A vintage photo of a train leaving the Gösser brewery. (Bräu Union)**

Above: Bräuerei Puntigamer in Puntigam, a district of Graz. (Bräu Union)

Below: Major breweries from the Bräu Union family are represented by Gösser's Spezial, Puntigamer's "das bierige Bier" lager, Schwechater Bier (a lager), Wieselburger Gold and Zipfer Märzen. (Bräu Union)

Bräuindustrie, or Steirerbräu, it was headquartered at Graz in the province of Steiermark (Styria), where commercial brewing had been ongoing since at least 1478. The major components of this consortium, both operating near Graz, were originally Reiningshaus and Puntigamer. They joined forces in 1943 after an air raid destroyed the

Reiningshaus Brewery and severely damaged Puntigamer. In 1947, brewing operations were consolidated in the latter brewery, but the two brands existed separately.

In 1977, the Gösser Bräuerei in Leoben joined Steirerbräu. One of Austria's largest breweries, Gösser was founded in 1860 by Max Kober on the site of a monastic brewery that had existed as early as 1459. In 1992, Steirerbräu was incorporated into Bräu Union in a hostile takeover, and in 2003 Bräu Union itself was absorbed into Heineken.

Today, Bräu Union offers a flagship pale lager in the 5 to 5.4 percent ABV range for each of each of its brands, including Gösser, Kaiser (brewed by Wieselburger), Puntigamer,

Reiningshaus, Schadminger, Schwechater, Wieselburger, and Zipfer. Other lagers include 4.5 percent ABV Gösser Dark, the 4.7 percent Kaiser Doppelmalz dunkel, 5.1 percent Mönchsbräu Märzen, 6 percent Puntigamer Winterbier Märzen, 5.4 percent Schwechater Zwickl, 5 percent Zipfer Märzen, and 5.4 percent Zipfer Urtyp.

With the Heineken acquisition of Bräu Union, the largest privately owned brewing company in Austria was the Stieglbräuerei in Salzburg, which has been in operation, "always using the best spring water from the local Untersberg mountain," since 1492. A company statement declares that "as the largest privately owned brewery in Austria, we have succeeded as an independent and self-sufficient family business. We wish to continue to produce Austria's favorite beer, which is why we are dedicated to consistent quality and in turn to the art of brewing at the highest level!"

In the Stiegl portfolio are 4.9 percent ABV Stiegl Pils and 4.9 percent Stiegl Märzen as well as several hefeweizens, including Stiegl Weizen Gold and Stiegl Weisse Naturtrüb, both at 5.1 percent. At the higher end of the ABV scale are a 7 percent bock beer called Original Stieglbock, and 9.2 percent Stiegl Double IPA.

Other independent breweries include the Vereinigte Karntner Bräuereien (aka Villacher

Bräuerei). Founded in 1738 in Villach, which is in the mountains midway between Innsbruck and Graz, it was rebuilt in 1858 by Johann Fischer. The portfolio includes Villacher Hausbier, Villacher Märzen, Villacher Zwickl, and Piestinger Lager, all 5 percent lagers.

The Johann Kühtreiber Brewery was founded in the city of Laa on the Thaya River in 1454. Kuehtreiber brews Hubertus Bräu,

**Above and below: Hofbräu Kaltenhausen in Hallein-Kaltenhausen near Salzburg was founded in 1475, declined over time and closed. However, it was recently reopened as a "speciality production centre, in which new beer recipes are constantly developed. [Kaltenhausen also] provides a forum for beer culture, that is popular amongst both professionals from the food industry and beer fans." (Bräu Union)**

**Above: Siegfried "Sigi" Menz (right), also known as "Mister Ottakring," is chairman of Getränkeindustrie Holding AG, while Christiane Wenckheim leads Ottakringer Bräuerei AG, the company's Vienna-based brewery subsidiary. (Ottakringer)**

Märzen and 7.5 percent Hubertus Festbock.

The Trumer Bräuerei in Salzburg traces its roots back to the brewery that was blessed by Archbishop Wolf Dietrich in 1601. The brewery was acquired in 1775 by hop merchant Josef Sigl, and it has been in the same family ever since, though the physical plant was destroyed during the Napoleonic wars and in a fire in 1917. The brewery was upgraded by Josef Sigl VII in the 1980s, and in 2003 he collaborated with Carlos Alvarez of the Gambrinus Company to build a second brewery in Berkeley, California, to brew the flagship Trumer Pils in the United States using ingredients shipped from Europe. The flagship brand, and the only beer brewed in Berkeley, is Trumer Pils, which is rated at 4.9 percent ABV.

In Vienna, Austria's capital and largest city, the last large brewery is Ottakringer Bräuerei, which was founded by Heinrich Plank under the name of Planksche Bräuerei in 1837 just as the lager revolution was about to change the course of brewing history. It was acquired in 1850 by a pair of cousins, Ignaz and Jakob Kuffner, and remained in the family until they were forced out by the Nazis because the Kuffners were Jewish. After World War II and a period of Russian control, the Kuffners were compensated for the confiscation, and the

whose labels feature the distinctive stag of Saint Hubert. In the seventh century, Hubertus had been encouraged to lead a saintly life after encountering a vision of a crucifix between the antlers of a stag. In the brewery's lineup are 5.1 percent ABV Hubertus

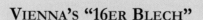

### VIENNA'S "16ER BLECH"

Ottakringer is well-known for its canned 5.4 percent ABV Vienna-style lager, which came to be known on the street as "16er Blech." The latter term means "sheet metal" but is also slang for "can," and Ottakringer is located in Vienna's 16er Bezirke (16th Distinct). Therefore, this Ottakringer product is the "16th District Tin Can." The company has embraced the name and uses it on the cans.

brewery was acquired by the Harmer family. It is now controlled by the holding company Getränkeindustrie Holding AG, which owns an interest in the Innstadt Bräuerei in Passau.

In addition to its flagship 5.4 percent ABV Vienna-style lager, known as 16er Blech (see sidebar) Ottakringer products include 5.2 percent Gold Fassl Dunkles, 4.8 percent Gold Fassl Pils, 5 percent Ottarocker, 5.1 percent, Ottakringer Helles, 5.2 percent Ottakringer Zwickl, 4.6 percent Gambrinus, and 7.6 percent Ottakringer Bock.

Vienna is also home to a growing number of inviting brewpubs, some of which brew ale as well as lager. The atmosphere ranges from that of a traditional American sports bar (although the "football" is not the same) to that of a typical Austro-German beer hall, and from the cozy and family friendly to the feel of an urban nightclub. Those which are farther from the center of the city, where space is not so much of a premium, have outdoor beer gardens. At press time, those in the city center included 1516, Bauern Bräu, Perchten Bräu, Schwarzenberg Stadtbräuerei, and Wieden Bräu. Plutzerbräu and Sieben Stern Bräu are in the museum district. Highlander, Lichtenthaler Bräu, Salm Bräu, Fischer Bräu, Grinzinger Bräu, Kadlez, Medl Bräu, and Schwarzer Rabe (Black Raven) are farther afield.

**Above: Massive steins are brandished by a pair of very serious Gösser fans. (Bräu Union)**

**Below: Enjoying the music and fun at the Stieglbräuerei Bräustüberl restaurant in Salzburg. (Stiegl)**

# Switzerland

Historically, brewing in Switzerland begins with the legend of St. Gall, an Irish monk whose name enters the literature as one of the many patron saints of brewing, who is said to have founded Switzerland's first brewery in the seventh century near present-day St. Gallen. The city still has a major regional brewery—Schützengarten. St. Gall may have first brewed beer at St. Gallen in the seventh century, but Bräuerei Schützengarten, the town brewery, certainly dates to 1779. Products include the 5.6 percent ABV ale Gallus 612, the 5 percent Schwarzer Bar (Black Bear) dark lager, 5.2 percent St. Gallen Klosterbräu lager, and 5 percent St. Gallen Landbier.

The Feldschlösschen brewery in Rheinfelden is Switzerland's largest. The name means "Little Castle on the Hill," and this is the appearance of the brewery building, originally built in 1876 when the brewery was founded by Theophil Roniger and Mathias Wüthrich. With a direct connection to the Swiss national railway by 1890, Feldschlösschen became the country's largest brewery and continued to grow during the twentieth century, prospering unlike breweries elsewhere in Europe, as Switzerland avoided both world wars.

In 1991 and 1996, Feldschlösschen expanded even further through the acquisition of the Cardinal brewery in Freiburg, followed by a merger with Hürlimann in Zurich. The latter was long famed for its enormously powerful 14 percent ABV

Below: The old Schützengarten delivery truck on a hill overlooking St. Gallen. (Schützengarten)

Above: Schützengarten's Lager Hell. (Schützengarten)

Below: The poster suggests that Rugenbräu Alkoholfrei is brewed for friends. (Rugenbräu)

Samichlaus doppelbock. At the time, these companies were two of Switzerland's largest and most well-established breweries. Cardinal had been founded in 1788 by François Piller, and Hürlimann was founded in 1836 by Albert Hürlimann, who was a world leader in the scientific study of yeast. It had gone though five generations of Hürlimann family management when it was sold. Another old-time Zurich brewer that faded from the scene at the end of the twentieth century was Löwenbräu Zurich—not to be confused with the more well-known Löwenbräu Brewery in Munich.

In 2000, the Feldschlösschen Group, incorporating Cardinal and Hürlimann, became part of the Carlsberg portfolio. Each of the three brand names continues to be used by the new management for several products, all of them lagers. Cardinal is represented in the current lineup by 4.8 percent ABV Cardinal Lager and 5.2 percent Cardinal Speciale. From Feldschlösschen, there is 4.8 per-

cent Feldschlösschen Original, 5.5 percent Feldschlösschen Dunkel Perle dark lager, and 7 percent Feldschlösschen Stark strong lager. Rounding out the list is 5 percent Hürlimann lager.

The Calanda-Haldengut Brewery in Chur, became the third largest brewery in Switzerland in 1990, when Calanda Bräu merged with Bräuerei Haldengut of Winterthur. Calanda's roots go back to the small brewery started in Chur in 1780 by Rageth Mathis. This company grew by acquisitions, acquiring the Aktienbräuerei Chur and the Rhatische Aktienbräuerei, while becoming Calanda Bräu. Haldengut, meanwhile, traces its heritage to the Ernst family brewery in Winterthur in the early nineteenth century, but Johann Georg Schoolhorn, who entered the picture in 1875, is remembered as the man who put the brewery on the map. In 1993, shortly after the Calanda-Haldengut

FÜR FREUNDE GEBRAUT.

MIT FREUDE GETRUNKEN.

RUGENBRÄU

ALKOHOLFREI

**Above: The beers from Abbaye de Saint Bon-Chien in Jura have attracted a great deal of attention within the rarified atmosphere of French haute cuisine.** (Bill Yenne photo)

**Below: Three of Switzerland's most recognized brands, Cardinal, Feldschlösschen, and Schützengarten.** (Courtesy of the breweries)

merger, the company became part of Heineken under the earlier name, Calanda Bräu. Calanda continues with its independent brand identity. Some of the beers in the all-lager portfolio marketed under the Calanda name are 5.2 percent ABV Calanda Edelbräu, 6 percent Calanda Meisterbräu, and 2.4 percent Calanda Panache.

Another Heineken venture into Switzerland came in 2008 with the acquisition of the Eichhof Brewery in Lucerne. Eichhof traces its roots to an 1834 brewery, acquired in 1878 by Traugott Spiess, and to the Bavarian Brewery, founded in 1890 by Heinrich Endemann. These two breweries merged in 1922 as the Vereinigten Luzerner Bräuereien (United Lucerne Breweries), which took on the Eichhof (Oak Court) name in 1960.

Rugenbräu, an independent Swiss brewing company which has recently moved into whiskey distilling, is headquartered in Matten bei Interlaken, on the Canton of Bern. The operation was started in 1866 by a politician named Christian Indermühle. In 1875, after he died, his sons, Carl and Albert, built a stone cellar in Rugen Mountain near Interlaken for the lagering of beer. In 1892, the property was acquired by the Bavarian brewmaster Joseph Hofweber, who two years previously had bought the brewery in the Schlöss Reichenbach in Zollikofen. After being known as J. Hofweber & Company for most of the twentieth century, the brewery was renamed as Rugenbräu in 1968. Still owned by the Hofweber family, Rugenbräu began limited bottlings of its whisky in 2008. The beer products, all lagers, include 5.2 percent ABV Rugenbräu Spezial Bier Hell and 4.8 percent Zwickel Bier.

After an overview of large, long-established breweries in the German part of Switzerland, we turn to a relatively new microbrewery in the French part of the country. In 1997, a viticulturist named Jerome Rebetez started a small artisanal brewery called Brasserie des Franches-Montagnes in Saignelegier, near Jura. The story is similar to that of many microbrewers in the United States, who started out as homebrewers during their student years, were noticed by critics, and discovered their careers. His Abbaye de Saint Bon-Chien is an 11 percent ABV specialty that matures in oak barrels for 12 months, and earned a mention in the *New York Times* as the best barley-wine [actually it is a bière de garde] in the world. *Beer Advocate* gives Abbaye de Saint Bon-Chien a "world class" rating of 95, while *RateBeer* gives it a 99. Other beers include 10.276 percent ABV La Cuvée Alex le Rouge, described as a Jurassian Imperial Stout; the 7 percent La Dragonne, a spiced honey beer; the 8 percent La Mandragore dark ale; the 6 percent La Meule golden ale; 5.5 percent La Salamandre, a witbier; and 7.5 percent La Torpille, a dark spiced ale with prunes.

Rebetez serves his beers at his brewery but also sells them to restaurants and the specialty beer market. As the press release reports, "The gastronomy industry has also found pleasure in products from the Brasserie des Franches Montagnes. More and more restaurant chefs use [our] beers in their meals to accompany meats, pastries, and desserts."

Above: The gleaming copper brew kettles in Feldschlösschen's 1876 brewing hall. (Feldschlösschen)

Below: A brewery wagon and team on the cobbled streets of Rheinfelden. (Feldschlösschen)

# The Netherlands

Walking the streets of any Dutch city or town, it is nearly impossible to avoid seeing the greens signs with the red star which proclaim the ubiquity of the country's leading brand. While a number of small craft breweries have come and gone in recent decades, Heineken dominates the Dutch market like no other single brand in any other country, with the exception of Guinness in Ireland. At the same time, Heineken is so widely known around the world that its origin in the Netherlands is often overlooked by consumers.

**Below: Enjoying a beer aboard an Amsterdam houseboat. (Grolsch)**

Heineken ranks number three in the world behind AB InBev and SABMiller, while the company's flagship brand has perhaps more global market visibility than any other. The company operates more than 100 breweries in more than 70 countries, including s'Hertogenbosch and Zouterwoude in the Netherlands. Today, approximately 25 million hectoliters of beer is produced in the Netherlands annually, roughly double the total in 1980. Worldwide, Heineken brews roughly 140 million hectoliters annually.

At home, the company dominates a domestic market which experienced many decades of consolidations in the twentieth century. In this environment, many middle-tier companies have been gradually acquired by international conglomerates, but others remain independent amid a rising tide of interest in specialty brewing that

**Above: Heineken's globally familiar flagship brand is five percent ABV, and 23 on the IBU scale. It is brewed in the Netherlands and under license at other located throughout the world. (Heineken)**

**The original Heineken flagship brewery in Amsterdam operated from 1867 to 1888, and is now a visitor center called The Heineken Experience. (Heineken)**

**Above: The tap handles inside the tasting room indicate what beers are available today at Koningshoeven. The portfolio includes a witte, a blond, and a bockbier, as well as a dubbel, tripel, a 10 percent ABV quadrupel, and a quadrupel aged in oak. (Koningshoeven)**

**Below: The bike rack at Koningshoeven. Many beer travelers tour the Netherlands by bicycle. (Koningshoeven)**

has also spawned a growing number of microbreweries.

In medieval times, the area that is now the Netherlands was an important brewing center known for its numerous small village breweries which produced dark top-fermented beers. One of the indigenous beer styles was oud bruin (old brown), a dark beer that is aged for up to a year. Even as lager has come to dominate the market, many larger brewers, including Heineken itself, still brew oud bruin.

The modern history of Netherlands brewing starts with Heineken, a story that began in 1863, when Gerard Adriaan Heineken noticed that lager beer was becoming more popular than the darker top-fermented beers brewed by most of the country's nearly 600 small brewing companies. An apocryphal tale relates that Gerhard told his mother of his idea that a good lager would answer her desire to rid Amsterdam's Sunday morning streets of wayward souls who had indulged too much in hard liquor the night before. Gerard proposed that he could provide a lighter alternative of a consistent quality, if she would only provide the money.

In 1864, after convincing his mother of the merits of his plan, he bought De Hooiberg (The Haystack), an old brewery and well-known Amsterdam establishment dating from 1592. In 1874 he expanded even further, with the opening of a brewery in Rotterdam. Having become one of his nation's most successful brewers, Heineken turned to exporting his beer to European markets.

At the turn of the twentieth century, Heineken's biggest competitors in the Netherlands were Amstel and Oranjeboom, located respectively in Amsterdam and Rotterdam, both of which still exist as brands. The Amstel Brewery was founded in 1870 by C.A. de Pesters and J.H. van Marwijk Kooy and named for the river that flows through Amsterdam. Oranjeboom (Orange Tree) dates back to 1671, and in the nineteenth century it was one of the first breweries in the Netherlands to produce lager.

The first half of the twentieth century, with the Great Depression sandwiched by world wars, was a period of economic hardship and decline in overall volume for Dutch brewers. The Dutch brewing companies that flourished were those who built a solid export business, and no one did this better than Heineken. Even before the 1920s, Gerard Heineken's son and heir, Henry Pierre Heineken, decided that the key to survival and further growth was expansion into markets outside Europe. This included the Netherlands East Indies (today Indonesia), Africa, the Caribbean, and the Far East. In 1933, soon after the repeal of Prohibition, Heineken became the first European beer imported into the United States.

As the Dutch market recovered after World War II, Heineken had a 25 percent market share in the Netherlands compared to 17 percent for Amstel. Total national production doubled to 2.2 million hectoliters in the 1950s and then increased five-fold by 1965. By the 1960s, the top tier of the 10 million hectoliter Dutch market was divided 31 and 19 percent respectively, although Heineken had a much larger share of the draft beer market in the country and a vastly stronger position overseas. Against this backdrop, Heineken and Amstel merged in 1968, although

Amstel still maintains an independent brand identification within the Heineken family and continued to operate from its own Amsterdam brewery until 1982. Meanwhile, Heineken terminated production at its own flagship brewery in Amsterdam in 1988, reopening it as a visitor center in 1991 and again in 2008 as a substantially revamped facility renamed the Heineken Experience. Heineken operations in Rotterdam were relocated to larger, more modern facilities in Zoeterwoude.

The second-largest Netherlands brewing company is Grolsch, founded in 1615 by Willem Neerfeldt in the town of Grol (now Groenlo). The company was acquired in 1895 by the de Groen family and became Koninklijk Grolsch (Royal Grolsch) through an official proclamation. Now brewing at a facility in Enschede, Grolsch was acquired by SABMiller in March 2008.

**Above: The Bavaria Brouwerij near Lieshout in North Brabant. (Bavaria)**

**Below: A memorable brewery promotion came when Bavaria packed the stands with women in orange (Netherlands' national color) during the Netherlands game at the 2010 World Cup. Bavaria was not an official sponsor, so FIFA complained of "ambush marketing." (Bavaria)**

**Above, left to right: Zatte, a tripel, the first (1985) and classic beer of Brouwerij 't IJ. Slightly sweet, dry finish, 8 percent ABV. Natte is the dubbel from the brewery. Smooth, dark and 6.5 percent. IJbok, the annual autumn bokbier is "dark and full-bodied," with 6.6 percent. Brewed for the "dark days of winter," the spicy IJndejaars changes annually and has an ABV of 9 percent ABV. (Brouwerij 't IJ)**

**Below: The terrace at Brouwerij 't IJ in Amsterdam, with its long tables and fresh beer is a popular destination for beer travelers from around the world who come here seeking good beer in a beautiful setting. (Brouwerij 't IJ)**

Several mid-sized brewing companies are located in the province of Limburg, including Alfa in Schinnen, Gulpener in Gulpen, and the Brand Brewery in the tiny hamlet of Wijlre. The Alfa Brewery was founded in 1870, while Gulpener dates to 1825. The Brand Brewery, acquired by the Brand family in 1871, traces its roots back to 1340, making it the oldest in the Netherlands. Two mid-sized breweries located in North Brabant are Dommelsche of Dommelen, founded in 1744, and the Bavaria Brewery in

Lieshout, which was founded in 1719 by Laurentius Moorees.

Gulpener is a family-owned brewery founded in 1825 by Laurent Smeets, a man who combined the profession of brewer with that of Justice of the Peace. His brewery, called De Gekroonde Leeuw (The Crowned Lion), was maintained by his sons and then in turn by their sons and a son-in-law named Rutten. After three generations, Gulpener passed into the possession of the Rutten family, who still manage the company.

Brand is now owned by Heineken, while Dommelsche is now part of the AB InBev portfolio. Another AB InBev acquisition is the Arcense Brewery in Arcen, which was renamed as Hertog Jan (Duke John) after its flagship brand. Jan I, known as Jan of Brabant, was a popular and fun-loving thirteenth-century monarch who was, and still is, celebrated in popular literature. After these acquisitions, Bavaria has the distinction of being the second-largest independent brewing company in the Netherlands after Heineken.

Since the 1980s, smaller breweries have revived older styles in the Netherlands. Meanwhile, as elsewhere in the world, new craft breweries are offering Dutch consumers more in terms of flavor and complexity. The newer Dutch brewers also pay homage to their neighbors, with both Belgian and German styles being found in the portfolios of the respective brewers. On the Belgian side, dubbels and tripels are frequently found as are witbiers. German-style lagers are often featured, and one might see a bock beer (bokbier) seasonally. Many brewers also brew porters and stouts, which originated as styles in Britain and are widely brewed in the United States. Recently, we have been seeing and enjoying the Dutch interpretation of the American interpretation of the British IPA.

In neighboring Belgium, there are several Trappist abbey breweries, and in the Netherlands there is one, the Abbey of Onze Lieve Vrouw van Koningshoeven near Tilburg. The Trappists at Koningshoeven began brewing in 1884, and today their beers are available on the commercial market under the name La Trappe.

One of the most famous of the new generation of Dutch microbreweries is Brouwerij 't Ij in Amsterdam. Founded in 1985 by musician Kaspar Peterson, it is named for the Ij (pronounced "eye"), the bay which fronts the city of Amsterdam. Located in the shadow of the picturesque De Gooyer windmill, the brewery maintains a pub and terrace that are a "must" destination for beer tourists visiting Amsterdam. Brouwerij 't Ij also bottles its popular products for retail sale.

Another Dutch craft brewery with a windmill is Brouwerij De Molen (Windmill Brewery) in Bodegraven, which is located inside the 1697 mill known as De Arkduif (The Dove of Noah's Ark). Founded in 2004 by Menno Olivier, late of Stadtsbrouwerij De Pelgrim in Rotterdam, De Molen has won gold and silver medals at numerous European beer festivals, and has been listed among *RateBeer*'s "Best Brewers in the World." Rated high on many lists is 10.2 percent ABV Hel & Verdoemnis (Hell & Damnation) Imperial Stout, self described by the brewery as its "best beer."

Above: De Molen offers several variations on its highly regarded Hel & Verdoemnis (Hell & Damnation) Imperial Stout. (De Molen)

Below: Brouwerij De Molen is located in a windmill in Bodegraven. (De Molen)

# Belgium

Below: In his paintings of sixteenth century folk life in what is now Belgium, Pieter Bruegel the Elder often included people enjoying beer. Stoneware carafes in the style in which the man on the left is pouring beer are still in use in Belgium today. (Author's collection)

Though it is a small nation and a relatively small-volume producer of beer, Belgium ranks as one of the world's most important brewing nations and an unprecedented experience for the beer traveler. Belgium is home to the mysterious and rare Westvleteren 12, a monastery-brewed ale that consistently appears near the top of nearly every serious list of the best ten beers in the world. Meanwhile, it is hard to find such a list on which Belgian brewers are not well represented.

In Belgium, whether in Flanders in the north or Francophone Wallonia in the south, the traveler may choose from a dizzying variety and level of quality that is almost unheard of elsewhere. The fact that every beer style, and many brands within the styles, have their own distinct glass is a subject of great fascination for the beer traveler when visiting a bar that has many beers on tap. Racks upon racks of differently shaped glassware are above, around, and behind the bar.

The beer traveler will also enjoy Belgium for its compactness. Many world-class artisanal breweries are located within a relatively short distance of one another, separated only by the picturesque Flemish and Wallonian landscape. Indeed, the level terrain allows many travelers to make their brewery tours by bicycle. In such a tour, the traveler will encounter many more breweries than we are able to mention in the brief overview that follows.

In medieval times, beer was first brewed in villages and on small farms, a tradition that still continues, although many family-owned breweries grew into substantial commercial companies in the nineteenth century. Until the eleventh century, commercial beer was the prerogative of the monasteries, six of which are still brewing commercial quantities of beer today. Later, brewing became a trade and brewers formed influential guilds to protect not only the producers but their customers.

The superb seventeenth-century guild house of Belgian brewers, which still stands on the spectacular Grand Place in Brussels, bears witness to the importance of the Belgian brewing industry. The Belgian Beer Board, the industry advocacy organization, is still headquartered in the building, whose windows look down upon the annual Belgian Beer Weekend, a festival held in the Grand Place each September. More than 400 Belgian brewers come to offer their wares. When the organiza-tion coined the phrase "Belgian Beer Paradise," it was an understatement. It is virtually impossible for any beer traveler to experience the entire scope of Belgian beer and brewing.

The overall size of the Belgian brewing industry declined in the mid-twentieth century but

**Above: The coach from Brouwerij Palm crosses the Grand Place, the central square in Brussels. Each year in September, the Grand Place is the scene of the Belgian Beer Weekend, one of Belgium's most internationally famous beer festivals.**
**(Bill Yenne photo)**

**Above: In addition to their beers, the monks of the Chimay Abbey also make cheese. (Manneken-Brussel Imports)**

**Below: Declared the "best beer in the world" in 1998, the flagship beer of Brouwerij Huyghe is still in the top tier of many lists. (Coaster from the author's collection)**

bounced back and has reached new levels in the twenty-first century. Annual production, which stood at 15.6 billion hectoliters in 1930, dipped to 10.3 in 1965 but rose to 14.6 in 1992 and to 18 billion in 2010. While a major part of this is due to lager brewing, the renewed global interest in the unique variety of Belgian beers had also played a role.

The world's largest brewing company, AB InBev, is nominally headquartered in Belgium, although it is a multinational company with many centers of gravity from New York to Sao Paulo, and it's traded on the New York Stock Exchange as well as Euronext. It evolved

from points on three continents. The Belgian point was Interbrew, formed in 1988 through a merger of Belgium's two largest breweries, Brouwerij Artois (famous for its Stella Artois lager) in Leuven and Piedboeuf in Jupille (famous for its Jupiler lager). Meanwhile, in Brazil in 1999, a company called AmBev was created by a consolidation of brewing companies of which the flagship lager was Brahma. Interbrew and AmBev merged in 2004 to become InBev, bought Anheuser-Busch in 2008, and became AB InBev.

Of the original Belgian components, the Jupiler Brewery was founded by the Piedboeuf family in 1853. Launched in 1966, Jupiler Lager became Belgium's largest-selling brand. The Artois Brewery dates to the House of Den Horen, an inn in Leuven which first brewed its own beer in 1366 and from 1537 sold its products to the University of Leuven (founded in 1425). Den Horen had many owners before Sebastien Artois bought it in 1717. He had been an apprentice there, earning the title of master brewer in 1708. Stella Artois (Star of Artois) was named after the Christmas star and was a "special occasion" beer launched in 1926. Loburg was launched in 1977 as a "premium" variation on Stella Artois Lager.

Today, the Belgian mainly-lager portfolio of AB InBev includes 5 percent ABV Stella

Artois, 3.4 percent Stella Artois Light, 5.7 percent Loburg, 5.2 percent Jupiler, and Piedboeuf Triple, which is not a tripel, but a 3.8 percent lager.

Another large lager consortium is Alken-Maes, formed in 1988 by a merger of the lager brewers Cristal-Alken of Alken and Maes of Kontich-Waarloos. In 2000, Alken-Maes bought Brasserie Ciney, located in the town of the same name, and was itself acquired by Scottish & Newcastle. In 2007, when Carlsberg and Heineken jointly acquired Scottish & Newcastle to split its assets, Heineken took Alken-Maes. The product line includes the long-time flagship lager, 5.2 percent ABV Maes Pils, as well as Grimbergen "abbey beers" and Mort Subite lambics. Also produced is 8.5 percent Hapkin, an ale originally brewed in the late twentieth century by Jackie Louwaege of Kortemark, said to have been invented in the Middle Ages by the monks at the Abbey of Ter Duinen at the request of the then Count of Flanders, Hapkin-with-the-Axe.

Many small- and medium-sized brewers still remain in towns and villages. Some have grown into large regional companies, but most remain committed to styles other than lager, and it is to those which turn our attention.

In defining Belgian beers generally, which is difficult because there are so many exceptions, we should say that Belgian beers are

## Belgian Beer Festivals

Belgium's highest profile beer festival is the **Belgian Beer Weekend**, held annually since 1997 in the Grand Place in Brussels during the first weekend of September. It is sponsored by the Knighthood of the Brewers' Mash Staff (the Belgian Brewers' Guild), which perpetuates the traditions and nobility of the brewer's trade as developed over the centuries. The guild also organizes a series of festivities every year in January or February to pay tribute to Gambrinus, the legendary "King of Beer."

Another weekend event on the Belgian beer calendar is the **Week-End Bières Spéciale**s that has been held since 1993, usually on the last weekend of February, in the village of Sohier, near Wellin in the Ardennes foothills. It is a small festival affording an opportunity to taste some extraordinary beers in a leisurely setting which browsing the offerings of breweriana dealers. Another small town that looms large on the Belgian beer calendar is Eizeringen, which hosts the **Nacht van de Grote Dorst** (**Night of Great Thirst**), aka the **International Geuze and Kriek Festival**, in even numbered years.

Other events in Eizeringen are the **Day of the Kriek** in June, the **Day of the Lambic** in December, and the Toer de Gueuze (Tour de Gueuze), a tour of Flemish breweries held during the spring in odd-numbered years.

In April, there is the **Zythos Beer Festival** (**ZBF**) at the Brabanthal in Leuven, which features more than 100 Belgian brewers, and is one of Belgium's larger festivals. In June, beer travelers make their way to the Groenplaats in Antwerp for the annual **Bierpassieweekend** (**Beer Passion Weekend**), which was created in 2000 by Ben Vinken, the editor & publisher of *Bierpassie/Bièrepassion Magazine*. In mid-August, it is time for the **Internationaal Streekbierenfestival** in Zwevegem. Rounding out the annual beer festival calendar is the **Kerstbier Festival** (**Christmas Beer Festival**), which is held annually in mid-December at the Heuvelhal in Essen.

Inset: This logo, both widely seen and appropriate in its message, may be used only by members of the Union of the Belgian Breweries. (UBB)

not as highly hopped with bittering hops as craft beers in the Anglo-American world, and that Belgian yeast strains provide a distinctive fruitiness and floral character. Unlike German beers, which adhere to the strict water-

Above: The old "brouwzaal" (brewing hall) at Brouwerij Lindemans, a well-known lambic brewery in Vlezenbeek. (Lindemans).

Below: This selection of popular Belgian lager brands includes Palm Speciale, brewed in Steenhuffel; Stella Artois and Stella Artois Light, brewed in Leuven; and Maes Pils, brewed in Alken. (Courtesy of the respective breweries)

than Anglo-American pale ales, usually straw colored or a little darker, and are less bitter and more delicate in their hop finish. When poured, they are distinctively topped with large, dense, long-lasting white heads and plenty of lacing on the glass.

The Belgian dark ales are brown or deep ruby red. The flavor and aroma is usually spicy or fruity and more complex, although the other characteristics of their pale cousins, especially the lack of bitterness, continues. When the ABV of either the pale or dark exceeds about 6 or 7 percent, they tend to be referred to as strong pales or strong darks. In recent years, as Belgian brewers have increased bittering hops in some variants for the American

hops-barley restrictions of the Reinheitsgebot, Belgian tradition allows for spices, fruit, added sugar, and plenty of experimentation.

We begin our overview of styles with the top-fermenting types and the Belgian "golden ales." Sometimes called Belgian pale ales, they are lighter in color

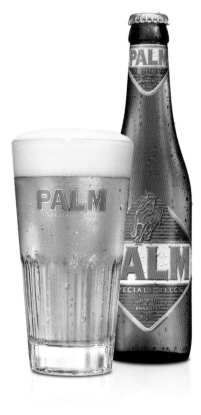

market, some commentators have used the term IPA to distinguish them from their traditional cousins. Belgian brewers also produce brown ales called oud bruin (old brown), which are close cousins to the oud bruins brewed across the border in the Netherlands.

Flanders red ales, which take their name from their reddish brown color, are typically more tart than other darker Belgian ales. They display a complexity that comes from extended aging in oak barrels and the blending of aged and young beer.

When it comes to identifying ABV, a traditional scale used in Belgium classifies a beer (of any color or shade) that is roughly between 6 and 9 percent ABV as a "dubbel," though it is not literally "double" the ABV of the "simple" variant. Meanwhile, the term "tripel" identifies a beer with an ABV between 9 and 12 percent. The terms originated in Trappist breweries to describe the use of double or triple the amount of malt in brewing relative to basic variant. Despite the higher alcohol content, other ingredients in the beer are also increased so that the overall effect is pleasingly balanced. Secular brewers have adopted these terms, and in recent years the term "quadrupel" had entered the lexicon.

A distinctive family of Belgian beers are the lambics,

a form of wheat beer which have long been made in the Senne Valley, south of Brussels. Until only recently, when experimentation began elsewhere, lambic beers were distinct from nearly all other beers in the world in that they are produced with neither lager yeast nor ale yeast. They are spontaneously fermented with wild yeast native only to that valley. Of the dozens of microorganisms that have been scientifically studied in connection with lambic beer, the defining yeast strain that causes the desired spontaneous fermentation is *Brettanomyces bruxellensis*.

Because lambic is an inherently sour beer, it was tradition-

Above: In Belgium, where every beer has its special glass, there is perhaps none more unique than that of Pauwel Kwak. (Bosteels)

The brewhouse at Brouwerij Bosteels in Buggenhout, where Kwak is brewed. (Bosteels)

**Above: Scaldis Ambrée is also marketed under the brand name Bush Ambrée. The beers of Brasserie Dubuisson Frères in Leuze-Pipaix use both brand names.** (Dubuisson)

**Below: A moody photo of a model train posed with Scaldis Ambrée in alternate glassware in a landscape familiar to the travelers visiting Belgium.** (Dubuisson)

ally flavored with tart, wild cherries and named kriek, after the Flemish word for the dark-red morello cherry. By the twentieth century, kriek lambics were being joined by framboise (or frambozen) lambics flavored with raspberries and cassis lambics flavored with black currents. By the 1990s, brewers were marketing lambics flavored with apples, apricots, grapes, plums, strawberries, bananas, and pineapples. Lambics made with a mixture of flavors and aged for several years are called gueuze. Lambics that are flavored with sugar, molasses, or other beer styles are called faro.

Belgium also has an indigenous unfiltered wheat beer. As German wheat beers are often called weiss beers, or "white" beers, so too are Belgian wheat beers. In Flemish, a white beer is a witbier. In Belgium's French-speaking south, the same beer is called bière blanche. The Belgian white has the cloudy appearance of a German hefeweizen, but is flavored with coriander and sometimes orange peel.

A more recent Belgian beer style is bière de champagne. The name applies to the fermentation method, as the beers themselves vary widely from golden to very dark. Some have added spices, but others do not. They are brewed mainly in Belgium, although breweries from Italy to the United States have been experimenting with the style. Some are aged in champagne barrels, and some Belgian brewers, Brouwerij Bosteels for example, have even sent the beer to be aged in caves in the Champagne region of France and put through the "méthode champenoise" to extract yeast from the bottles. The resulting ABV typically exceeds 10 percent.

Brouwerij Bosteels was founded in 1791 in Buggenhout by Evarist Bosteels and has been in the family for seven generations. Their bière de champagne, DeuS Brut des Flandres, is aged for two months in Flanders and for nine months in France. The result is rated at 11.5 percent ABV. Bosteels's 8.4 percent Tripel Karmeliet golden ale based on a 1679 recipe from a Carmelite convent in Dendermonde. To meet demand, the beer is also brewed by Brouwerij Van Steenberge.

There is a story behind serving Bosteels' 8.4 percent ABV Kwak in its proper glass. As the story goes, there was an eighteenth-century brewer and innkeeper named Pauwel Kwak who, among other things, supplied beer to coachmen who dropped passengers at his inn, named De Hoorn. He designed a glass, a shorter version of an old English "yard," that would sit in a bracket next to a coachman's seat. In the 1980s, Bosteels relaunched the glass, which is sold with a wooden bracket for consumers who don't have coaches.

Brasserie Dubuisson Frères was founded in 1769 by Joseph Leroy as a farmhouse brewery near Pipaix, and it's still in the family. In 1931, the Dubuisson brothers Alfred and Amédée, Leroy's descendants, decided to turn the operation into a commercial brewery. They are known today for their 7 percent ABV Cuvée Des Trolls golden ale as well as for their interchangeable Bush and Scaldis brands. These include Bush Ambrée (aka Scaldis Ambrée) and Bush De Noël (aka Scaldis Noël), both at 12 percent ABV, and 10.5 percent Bush (aka Scaldis) Blonde Triple.

Brouwerij De Koninck in Antwerp is another brewery that was in the same family for almost two centuries, although it became an autonomous part of the Duvel Moortgat Group in 2010. It dates to 1827, when Joseph Henricus

De Koninck bought an inn De Plaisante Hof (The Pleasant Courtyard) across the street from a not-so-pleasant gallows. After he died, his wife married Johannes Vervliet. He started a brewery named De Hand (The Hand) on the site in 1833, which employed Joseph's son, Carolus De Koninck, who took over the business when Vervliet died. The hand still features on the company logo. A later partner in the business, Florent Van Bauwel, took over in the twentieth century, and it remained in his family until 2010. Beers included 5 percent ABV De Koninck, 8 percent Triple d'Anvers, and the seasonal 6.5 percent Winterkoninck.

Brouwerij Duvel Moortgat was founded in Breendonck in 1871 by Jan-Leonard Moortgat, and although it is publicly traded, it is still controlled by the family. A modestly-sized family business

Left: Cuvée Des Trolls is made with dried orange peel. (Dubuisson)

Below: The 8 percent ABV Tripel from Brouwerij De Koninck in Antwerp. (De Koninck)

This page, counter clockwise from above: The brewhouse at Brouwerij Duvel Moortgat in Breendonck; Emile and Leon Moortgat, grandsons of the founder, who ran the business in the mid-twentieth century; and Duvel, the quintessential Belgian golden ale. (Duvel Moortgat)

for a century, the company benefited from entering into a deal in the 1970s to bottle and distribute Tuborg. From this relationship, the Moortgat family was able to build a substantial distribution network for their flagship golden ale, Duvel.

This beer originated after World War I as Victory Ale, the English name an homage to the British who sacrificed much for Belgium in the war and a yeast strain that Albert Moortgat brought back from Scotland. At a tasting, however, a shoemaker named Van De Wouwer quipped that the 8.5 percent ABV beer was *"nen echten Duvel"* or "a real Devil." Having established their international distri-

bution network, Duvel Moortgat undertook a program of acquisitions in the twenty-first century.

Having been an early investor in the Brewery Ommegang craft brewery in Cooperstown, New York, they bought the brewery outright in 2003. This was followed by the acquisition of Brasserie d'Achouffe in 2006, Brouwerij Liefmans in 2008, and De Koninck in 2010. In addition to the flagship Duvel, other beers brewed at Breendonck include 9.5 percent ABV Duvel Tripel Hop and two lagers, 5 percent Vedett and 5.5 percent Bel Pils. Since 1963, Moortgat has had a license to brew the abbey beers developed by the monks at the Maredsous Abbey.

Brasserie d'Achouffe at Achouffe near Bastogne in the Ardennes highlands was founded in 1982 by homebrewers Pierre Gobron and Christian Bauweraerts. The labels feature the "chouffes," the legendary mute gnomes who found that drinking the water of the River Decrogne gave them the ability to speak. According to the story told by Gobron and Bauweraerts, in the brewery's early days the brewers left the raw materials for their beer in the brewhouse overnight, and the chouffes did all the brewing. The brewery, owned by Duvel Moortgat since 2006, has exported to North America since 1988 and has a string following there. The product portfolio is headed by 8 percent ABV La

Chouffe, an unfiltered bottle-conditioned beer which is "pleasantly fruity" and spiced with coriander. Other products include 8.5 percent McChouffe, a brown Scotch ale; 9 percent Chouffe Houblon, a heavily hopped golden ale, described as "a marriage between an Imperial IPA and a Belgian tripel"; and 10 percent N'ice Chouffe, a dark Christmas ale.

Brouwerij Huyghe was founded in 1906 by Leon Huyghe in city of Melle, near Ghent, on a site where brewing had been ongoing since 1654. The brewery has been substantially upgraded several times between 1939 and 2011. In 1988, Huyghe launched its now-legendary flagship beer, Delirium Tremens. An 8.5 percent ABV golden ale, it received a gold medal at the World Beer Championships in Chicago in 1998 at which it was called the best beer in the world. To celebrate the bicentennial of the French Revolution in 1989, the 8.5 percent "topical beer" La Guillotine was launched. The 10 percent Christmas beer Delirium Noël was added to the line in 2000. In 2013, Huyghe unveiled Deliria, the "sister of the world famous Delirium Tremens. The new beer—brewed for and by women—is not a typical sweet beer, but a heavy (8.5 percent) blonde Belgian ale, with fruity notes of apple and chardonnay grapes." In response to a Facebook, website, and newspaper call for entrants,

about 65 women from five countries applied to brew the beer, and ten were selected.

Family-owned since 1801, Palm Breweries traces its roots to the pub known as De Hoorn (The Horn) which existed in the village of Steenhuffel in 1597, though the brewery now on the site dates to 1747. The golden ale Palm Royale, now rated at 7.5 percent ABV, has been brewed since that date. The brewery was acquired in 1801 by Jan Baptist De Mesmaecker, whose descendant Henriette De Mesmaecker married Arthur Van Roy in 1908. The brewery continued to operate as De Hoorn, becoming Palm in 1975. After the ups and downs of the market and two world wars, the brewery celebrated its bicentennial in 1947 with Palm Dobbel, a dark 6 percent ABV which is still part of the portfolio. There

Above: Enjoying Maredsous at an outdoor cafe. (Duvel Moortgat)

Below: The Grand Cru of Brouwerij Rodenbach. (Rodenbach)

**Above: Brasserie d'Achouffe in the town of the same name, and the iconic "chouffe" (troll) that graces the label.** (d'Achouffe)

**Below: The author with d'Achouffe co-founder Chris Bauweraerts.** (Author's collection)

was substantial expansion in both plant and market presence during the latter quarter of the twentieth century, and in 1998 it took over the Rodenbach family brewery in Roeselare. In 2001, Palm acquired Brouwerij De Gouden Boom in Brugge, where brewing had taken place since 1455, which is famous for its own line as well as Steenbrugge abbey beers. Brewing operations were relocated to Steenhuffel in 2004.

The Rodenbach brothers, Pedro, Alexander, Ferdinand, and Constantijn, soldiers in Napoleon's army, acquired their brewery in 1821. Pedro bought out his brothers in 1836, and his wife, Regina Wauters, operated the brewery while Pedro pursued a military career. The controlling interest in the brewery was in the family until Palm acquired the business in 1998. Through the years, Rodenbach provided yeast for other breweries, from the De Dolle microbrewery in Esen to the legendary Westvleteren mon-

astery. The flagship beer is the highly regarded Rodenbach Grand Cru, a 6 percent ABV Red Ale.

De Halve Maan (The Half Moon) is now the only family brewery left in the center of the historic Flemish city of Brugge. In 1564, a brewery called Die Maene (The Moon) was operating near the present site. De Halve Maan was started in 1856 by Leon Maes, who is known in the family as Henri I. The business passed to Henri Maes II in 1867 and to Henri III in 1919. He learned lager brewing techniques in Germany and brought them to Brugge in 1928. In the 1950s, the contributions of Henry IV included expansion of a popular home delivery network. In 1981, Henry IV's daughter, Véronique, introduced the current flagship product, Straffe Hendrik (Strong Henri), a 6 percent golden ale. From 1988 to 2002, the beer was produced in commercial quantities by Brouwerij Riva. During those years, the brewery "opened its doors to the public. The house brewery was born. The former bottling devices and maltery were arranged into cozy dining rooms." In 2005, Xavier Vanneste, Véronique's son, revived commercial brewery activities after "a thorough renovation and modernization." Brewed onsite today are Straffe Hendrik, 6 percent ABV Brugse Zot golden ale, and the 7.5 present Brugse Zot Dubbel.

## Lambic Brewers

Among the breweries specializing in lambic beers, one of the most highly regarded is Brouwerij Boon in Lembeek, whose roots go back to 1680. In 1860, Louis Paul was brewing lambic on the site, and he began bottling gueuze lambic in 1875. The brewery was taken over in 1975 by Frank Boon, one of the pioneers in maintaining authentic lambic styles.

The story of Brouwerij Liefmans dates back to 1679, when Jacobus Liefmans started his brewery on the banks of the River Scheld in Oudenaarde. Kriek lambic brewing began around 1900, when "Liefmans started filling maturation tanks with black cherries, on a small scale. Local farmers brought their excess crops of black cherries to Liefmans and swapped them for beer." Current beers include 4 percent ABV Frambozenbier, 6 percent Kriekbier, and 8 percent Goudenband.

Brouwerij Lindemans is family-owned and operated by Dirk and Geert Lindemans. Located in the village of Vlezenbeek in Flemish Brabant southwest of Brussels, it was founded in 1822 by Frans Lindemans as a farmhouse brewery, but the family stopped farming and turned to full-scale commercial brewing in 1930, with a major brewery expansion coming in the 1990s.

Originally, Lindemans produced gueuze and kriek, with framboise added to the line in 1980, followed by cassis in 1986. In 1985, Michael Jackson called Lindemans' 4 percent ABV Kriek one of the five best beers of the world. Current

**Above: The Mort Subite cafe in Brussels, named for the "sudden death" endings of lunch hour card games, became the namesake for a family of lambics that are served here. (Bill Yenne photo)**

**Above: Tap handles inside the legendary Mort Subite cafe in Brussels.** (Bill Yenne photo)

**Below: Kriek (cherry) lambics from Brasserie Belle-Vue in Sint-Pieters-Leeuw, Brouwerij Boon in Lembeek, Brouwerij Mort Subite in Asse-Kobbegem, and Brouwerij Lindemans in Vlezenbeek.** (Courtesy of the breweries)

ABV ratings for the other beers include Lindemans Framboise at 2.5 percent, Lindemans Cassis at 3.5 percent, Lindemans Gueuze at 4 percent, and Lindemans Kriek Cuvée René at 6 percent.

Lambics are also produced by Brasserie Lefèbvre in Quenast,

which is better known for its Floreffe line of abbey beers.

Brasserie Belle-Vue was founded in 1913 by Philémon Vandenstock, acquiring its name in 1927 when he started operating the Café Belle-Vue in Anderlecht. He developed the brewery with his son Constant and son-in-law Octave Collin until his death in 1945. After World War II, the brewery became Belgium's leading gueuze producer, while becoming known for sweet, less tart, fruit lambics. The original Brussels brewery on the Willebroek canal was closed after Belle-Vue was acquired by Interbrew (AB InBev since 2008) in 1991, and production was moved to Sint-Pieters-Leeuw, southwest of Brussels. Products include Belle-Vue Gueuze, Belle-Vue Kriek, Gueuze De Koninck, all at 5.2

percent ABV, and Belle-Vue Extra Framboise at 2.9 percent.

A lambic name familiar to travelers in Brussels, though it is not brewed there, is Mort Subite. It is familiar from the name of the nineteenth-century café in the city center called A La Mort Subite—a destination of beer travelers—where it is served and from which it took its name. The term "mort subite," meaning "sudden death," comes not from the beer but from patrons abruptly ending lunchtime card games in order to get back to work. Actual brewing takes place at Heineken-owned Brouwerij Mort Subite, formerly Brouwerij De Keersmaeker, in Asse-Kobbegem, northwest of Brussels. Beers include Mort Subite Blanche, Mort Subite Framboise, Mort Subite Gueuze, and Mort Subite Kriek, all rated at 4.5 percent ABV, and 7 percent Mort Subite Oude Gueuze.

Brasserie-Brouwerij Cantillon was started in 1900 in the Anderlecht district of Brussels by Paul Cantillon and his wife, Marie Troch, both brewers, and is still family owned. One of the most highly regarded lambic brewers internationally, Cantillon is the last lambic brewery left in the Belgian capital, and it has used all-organic ingredients since 1999. Among Cantillon's most internationally acclaimed beers are Cantillon Fou' Foune, Cantillon Lou Pepe Kriek, Cantillon Lou Pepe Framboise, Cantillon Lou Pepe Gueuze, and Cantillon Gueuze 100 percent Lambic, all at 5 percent ABV, and Cantillon Saint Lamvinus at 6 percent.

**Above: Liefmans' highly regarded Cuvée Brut Kriek. (Liefmans)**

**Below: The spectrum of typical lambic styles includes Framboise, Gueuze, Faro, Cassis and Pêcheresse (Peach). (Lindemans via Merchant du Vin)**

Above: **Blanche de Namur from Brasserie du Bocq in Purnode, in the valley of the river Bocq.** (Bocq)

Below and right: **The classic witbier and Speciale Golden Wheat from Brouwerij van Hoegaarden in the town of the same name.** (Hoegaarden)

## BREWERS OF WITBIER (BIÈRE BLANCHE)

As might be expected of a wheat beer, witbier is indigenous to the wheat-producing open country in the Flemish-speaking regions of Belgium. Legend holds that witbier originated in Hoegaarden in Brabant, but it may have been introduced from Germany, although it is known that Monks were brewing witbier by 1445. It was popular throughout Belgium for five centuries, and once there were more than a dozen witbier breweries around Hoegaarden. After the middle of the twentieth century, the style was losing popularity, and one by one they closed. When Louis Tomsin closed his brewery in 1957, there were none, although Peetermans's witbier con-

tinued to be brewed by Artois in Leuven. In 1967, Pierre Celis, a Hoegaarden native who once worked for Tomsin, decided to revive the style and started his Brouwerij De Sluis. The popularity returned, and Celis was successful in the witbier revival. He bought a new production facility and launched his Oud Hoegaards (later called Hoegaarden).

After a fire in 1985, Celis received a loan from Artois to rebuild, and their involvement led to Celis's decision to sell them the brewery and his Hoegaarden brand. In 1992, he established his Celis Brewery in Austin, Texas, and continued brewing with the original Hoegaarden recipe, calling his beer Celis White. This brewery was sold to Miller Brewing and closed in 2000, though the brand was brewed for a time by the Michigan Brewing Company. Pierre Celis died in 2011, and in 2012 his daughter, Christine Celis, bought back the Celis brand name and announced plans for renewed brewing in Texas.

Meanwhile, Interbrew continued to brew the Hoegaarden brand in Hoegaarden, although plans were announced

in 2006 to close the brewery and consolidate production to Jupille. A firestorm of international opposition led to cancellation of these plans and an investment in the upgrade of Brouwerij van Hoegaarden to continue brewing the iconic 4.9 percent ABV Hoegaarden Witbier, the 8.5 percent Hoegaarden Grand Cru, and other products.

Hoegaarden may be the definitive witbier, and Pierre Celis the singular person in its history, but there are a number of other practitioners of the art, many of then in the United States.

A good example in Belgium is Brasserie du Bocq, founded in 1858 by Martin Belot in the valley of the River Bocq in Purnode. Since the revival of interest in witbier, the Bocq line has included Blanche d'Ardenne, Blanche De Namur, and Blanche Des Moines, each rated in the 4.3 to 4.5 percent ABV range, as well as Agrumbocq and Applebocq, both rated at 3.1 percent. Bocq also produces dark ales and lambics.

Another well-known witbier is Dentergems Wit, which originated with Brouwerij Riva in Dentergem but was later brewed by Liefmans. When Liefmans was sold to Duvel Moortgat in 2008, the Dentergems brand went to Brouwerij Het Anker in Mechelen. Acquired by Louis Van Breedam in 1872, Het Anker dates back to 1471 and the brewery of a Beguinage, or community of lay Christian women. It was later known for its 11 percent ABV Gouden Carolus D'Or (Grand Cru Of The Emperor), named after the Holy Roman Emperor, Charles V, who grew up in Mechelen. *Beer Advocate* reports that the 5 percent ABV Dentergems Wit is now brewed at both Het Anker and Riva.

**Above: Blanche de Bruxelles from Brasserie Lefèbvre in Quenast.** (Lefèbvre)

**Right: Brugs Witbier from Brouwerij Alken-Maes in Alken.** (Alken Maes)

**Left: Steenbrugge Wit from Brouwerij Palm in Steenhuffel.** (Palm)

Above: St. Bernardus Abt has 12 percent ABV. (St. Bernardus)

Below: Foreffe Dubbel has 6.3 percent ABV. (Lefèbvre)

## ABBEY BREWERS

Unlike Trappist beers, abbey beers (bières d'abbaye or abdijbier) are not necessarily brewed in an abbey nor by monks. Ideally, abbey beers are brewed from a recipe that originated at a real abbey and was licensed to a commercial brewery in exchange for royalties. However, the term came to be applied to commercial beers named for long-abandoned abbeys, saints not associated with actual abbeys, or entirely fictitious abbeys.

In 1999, the Union of Belgian Brewers introduced a Certified Belgian Abbey Beer (Erkend Belgisch Abdijbier) logo to be used on beers that were actually licensed by an abbey. Two well-known brands, Affligem and Maredsous, are licensed to brew the recipes of existing, active Benedictine abbeys. Brouwerij Affligem, originally the Op-Ale brewery in Opwijk, is officially licensed by the Benedictines at Affligem Abbey, which has existed since 1086 and is the Primaria Brabantiae, or principal monastery, in Brabant. The brewery is now owned by Heineken. The beers include 6.8 percent ABV Affligem Blonde, 6.8 percent Affligem Patersvat, 7 percent Affligem Dubbel, and 9.5 percent Affligem Tripel.

Maredsous Abbey is a religious community at Denée near Namur in Belgium that was founded by the Order of Saint Benedict in 1872. The monks of Maredsous are well known for the variety of cheeses that they make on premises, but their beer has been brewed for the commercial market by Duvel Moortgat since 1963. These include 6 percent Maredsous 6 Blonde, 8 percent Maredsous 8 Brune (dubbel), and 10 percent Maredsous 10 Triple.

Two other beers, Grimbergen and Floreffe, are named for abbeys established by St. Norbert of Xanten in the early twelfth century and occupied by the Premonstratensian (Norbertine) canons, who were long ago known for their beer. Both monasteries were disbanded at the end of the eighteenth century in the wake of the French Revolution and never reestablished.

Grimbergen takes its name from Grimbergen Abbey, built in the town of the same name in 1128. Brewing takes place at Brouwerij Alken-Maes in Alken. Beers include 6 percent ABV Grimbergen Blanche (witbier), 6.7 percent Grimbergen Blonde, 6.5 percent Grimbergen Dubbel, 9 percent Grimbergen Tripel, and 10 percent Grimbergen Optimo Bruno.

The Abbey of Floreffe was founded in the town of the same name in 1121 and served as the "mother house" for other Premonstratensian abbeys in the region. A brewery established here in the mid-thirteenth century was abandoned by the monks in 1794, restored in 1960, and operated commercially by Het Anker until 1983. At that time, brewing was taken over by Brasserie Lefèbvre in Rebecq-Quenast, founded in 1876 by Jules Lefèbvre. The beers include 4.5 percent ABV Abbaye De Floreffe Blanche, 6.3 percent Abbaye De Floreffe Blonde,

6.3 percent Abbaye De Floreffe Double, 7.5 percent Abbaye De Floreffe Triple, 8 percent Barbar (Belgian Honey Ale), a family of 3.5 percent Blanche De Bruxelles lambics, and 5 percent Manneken Pils.

A third Premonstratensian monastery, Abbaye Notre Dame de Leffe, near Dinant on the banks of the River Meuse, was founded in 1152 under the auspices of Floreffe and is still an active religious community. It survived the upheaval of the French Revolution period, but gradually the property

**Above: The official logo assigned to authentic abbey breweries by the Union of Belgian Brewers. (UBB)**

**Below: Abbey style beers from Grimbergen (Dubbel and Tripel) and Affligem (Blond and Tripel).** (Courtesy of the breweries)

**Above: Floreffe Abbey Tripel from Brasserie Lefèbvre. (Lefèbvre)**

**Below: Leffe Blonde and Leffe 9 (9 percent ABV) from Abbaye de Leffe in Dinant. (Leffe)**

was sold off, and the last monk died in 1844. The abbey was revived after 1903 by Premonstratensians that had been expelled from their abbey at Frigolet in France. Meanwhile, the first reference to a brewery at Leffe was in 1240, and this tradition was finally revived in 1952 when Father Abbot Nys brought in Albert Lootvoet, a brewer from Overijse, to run the brewery. With extensive investment being necessary to meet increasing consumer demand, the abbey decided in 1977 to license outside production at Brasserie Artois, now part of AB InBev. Leffe products include 6.6 percent ABV Leffe Blonde, 6.5 percent

Leffe Brune, 8.5 percent Leffe Tripel, and 8.2 percent Leffe Vieille Cuvée.

Sometimes referred to as an abbey, Priorij Corsendonk in Oud Turnhout was actually a priory, led by a prior, who is of lesser rank than an abbot. Coincidentally, it was founded in 1393 through the philanthropy of Maria van Gelre, youngest daughter of Jan III, Duke of Brabant. Jan I—Jan Primus, Duke of Brabant—is today better known as Gambrinus, the legendary king of beer. The monks at Corsendonk started their brewery around 1400 and brewed until the priory was closed in 1784. In 1906, Antonius Keersmaekers founded a secular brewery to revive the monastic brewing traditions. His award-winning beers included Agnus Dei (Lamb of God), a tripel described as "a light bodied, high fermentation brew of delicate palate." His brewery closed in 1953, but Corsendonk beer continues to be produced by Brasserie Du Bocq in Purnode. Beginning in 1968, the multiple buildings on the priory grounds were restored, and today Priorij Corsendonk is a luxury hotel and conference center unrelated to beer and brewing. In addition to 7.5 percent ABV Corsendonk Agnus, the beers include 3.3 percent

Corsendonk Apple White, 7.8 percent Corsendonk Summum Goud, and 8.5 percent Corsendonk Christmas Ale.

Brouwerij St. Bernardus, founded in 1946 in Watou, is a secular brewery associated over the years with two Trappist monasteries. As the company fact sheet notes, "one gave us our name, the other gave us our beer." The former is the community of Trappist monks from Mont des Cats in Godewaersvelde who left France in the nineteenth century to join an existing Trappist community near Westvleteren, which served as their as their Refuge Notre Dame de St. Bernard, but who returned to France in 1934. The latter is Trappist Abbey of St. Sixtus of Westvleteren, which is discussed in the following section on Trappist beers. After World War II, St. Sixtus decided to stop selling its

beer to the public and licensed its recipes to the lay brewery, just as the St. Bernard monks had licensed their famous cheese-making traditions to outside cheese makers. As the brewery notes, "We started selling the Trappist beers under the brand names Trappist Westvleteren, St. Sixtus or even

**Above:** The hop arbor at the entrance to Brouwerij St. Bernardus in Watou. (St. Bernardus)

**Below:** The brewhouse at Brasserie St. Feuillien Brewery in Le Roeulx. (St. Feuillien)

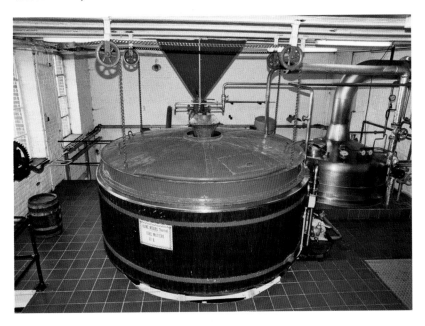

**This page: A visit to Brasserie St. Feuillien. Above is the old brick brewery building in Le Roeulx, while below are bottles in various sizes from the St. Feuillien product portfolio.**
(St. Feuillien)

later Sixtus. During a period of 46 years we brewed and commercialized the beers, while the monks continued to brew for themselves and for sales to three pubs in the neighborhood. . . . In 1992 the license came to an end and since then we are brewing the same beers, with the same recipes, but under a different brand name: St. Bernardus." These are 6.7 percent ABV St. Bernardus Pater 6, 8 percent St. Bernardus Prior, 10 percent St. Bernardus Abt 12, 5.5 percent St. Bernardus Witbier, and 10 percent St. Bernardus Christmas Ale.

The St. Feuillien Brewery in Le Roeulx in Hainaut Province was founded in 1873 by the Friard family. The name is said to be derived from an abbey established in 1125 on the site of the grave of the seventh-century Irish monk St. Faelan (in French, Feuillien). However, he is believed to have been buried around 655 at a monastery he founded at Fosses-la-Ville in the province of Namur, some distance southeast of Le Roeulx. The portfolio includes St. Feuillien Blonde and Brune, both at 7.5 percent ABV, as well as 9.5 percent St. Feuillien Grand Cru and 9 percent St. Feuillien Cuvée De Noël.

## TRAPPIST BREWERS

One of the most unique brewing traditions in the world is that of the monasteries of the Cistercian Order of the Strict Observance, better known as the Trappists. Their beer has become extremely popular, highly sought, and represents a dynamic line of continuity in the Belgian brewing heritage. For centuries monasteries were centers of literacy and study, art and science, and then evolved as brewing centers as well. They were also a refuge and resting place for pilgrims traveling to shrines or the Holy Land. Pilgrims who stayed for a night or two needed food and drink, so the monasteries brewed their own beer, made their own cheeses, and baked bread. Some still do.

Today, six of the world's eight Trappist breweries are in Belgium. These are Achelse Kluis (Achel), Chimay, Orval, Rochefort, Westmalle, and Westvleteren. The others are Koningshoeven in the Netherlands and Engelszell in Austria, where brewing resumed in 2012. Another Trappist monastery, Mariawald in Germany, recently ceased brewing. In 1997, the Trappist monasteries formed the International Trappist Association (Internationale Vereniging Trappist or L'Association Internationale Trappiste) to protect the "Trappist" trademarks from infringement in the marketing of beer, bread,

cheese, and many other products. These products are entitled to carry a logo identifying them as an "Authentic Trappist Product" only if produced by the monks or under their direct supervision.

Two of the Trappist abbeys in Belgium, Chimay and Westmalle, have an annual output in excess of 100,000 hectoliters. Orval brews about half this volume, and the others produce less than 20,000 hectoliters.

Abbaye Notre Dame de Scourmont near Chimay has brewed since 1863, producing three ales under the familiar Chimay label. Chimay Première,

**This page: The Chimay lineup, and Abbaye Notre Dame de Scourmont near Chimay, where the beer is brewed. (Manneken-Brussel Imports)**

This page, Orval Abbey's single product, and the lineups from the Rochefort and Westmalle abbeys. (Merchant du Vin)

copper-colored with a red label, has been produced since the nineteenth century and is rated at 7 percent ABV. Chimay Cinc Cents (500), with its white label, is an 8 percent tripel developed by the legendary brewmaster Father Théodore and introduced in 1966. Chimay Grande Reserve, with the blue label, is a 9 percent dark ale developed by Father Théodore in 1948.

The Abdij de Trappisten van Westmalle in Flanders has been brewing since 1836. They brew two bottle conditioned ales, the dark reddish-brown 7 percent ABV Dubbel and the clear golden-yellow 9.5 percent Tripel.

The Abdij der Sint-Benedictus de Achelse Kluis is the newest

of the Belgian Trappist breweries, having resumed the tradition with beer first marketed in 2001. The abbey was founded in 1648, destroyed in the French revolution, restored by monks from Westmalle in 1844, and restarted as an independent monastery in 1871. Brewing continued during this time, but the brewhouse was looted under the German occupation in 1917. Starting in 1998, Trappists from Westmalle and Rochefort helped rebuild the brewery.

The Abbaye Notre-Dame de St. Rémy, near the town of Rochefort, is the longest brewing monastic brewer in Belgium, having been active since 1595. Their portfolio includes the dark red-

dish 7.5 percent ABV Rochefort 6, with a red cap; the golden 9.2 percent Rochefort 8, with a green cap; and the dark 11.3 percent Rochefort 10, which has a blue cap. Of these, Rochefort 8 is the primary product, with Rochefort 6 produced only occasionally.

The Abbaye Notre-Dame d'Orval had an active brewing operation by 1628, which was interrupted by the French Revolution, and the present brewery was built in 1931. The abbey itself was not rebuilt until around 1926. Only one beer, a 6.2 percent ABV ale made with *Brettanomyces* yeast, is sold, although the monks produce the 3.5 percent Petite Orval for their own use.

De Sint-Sixtusabdij van Westvleteren (the Abbey of St. Sixtus of Westvleteren) was founded in 1831, with brewing begun in 1838. Between 1931 and 1946, and since 1992, the brewery has offered its beer for sale on a limited basis. In the interim, the products were licensed to Brouwerij St. Bernardus in Watou. Today, the Westvleteren Abbey brews one of best beers in the world and markets it under one of the most counterintuitive marketing schemes imaginable. Except under certain rare and limited circumstances, the beer is sold only at a small visitor center across the road from the brewery or at the brewery itself. Sales at the latter are done only by a reservation made by telephoning a number that

accepts calls only once a week. Beer will not be sold to the same customer twice in 60 days, and each customer must agree that the beer will not be resold to a third party. The three beers which are produced on a rotating basis are 5.8 percent ABV Westvleteren Blond, with a green cap, which was introduced in 1999 to supersede an earlier 6.2 percent beer; 8 percent Westvleteren 8, with a blue cap; and 10.2 percent Westvleteren 12, with a yellow cap.

Despite the fact that Westvleteren 12 is widely regarded as one of the world's greatest beers, the Trappists at the abbey have refused to consider increasing capacity to meet avaricious global demand.

As the father abbot said on the occasion of the opening of a new brewhouse some years back, "We have to live 'from' and 'with' our brewery. But we do not live 'for' our brewery. This must be strange for business people and difficult to understand that we do not exploit our commercial assets as much as we can. We are no brewers. We are monks. We brew beer to be able to afford being monks."

Above: The brewing hall at the Abbey of Notre-Dame de St. Rémy, near the town of Rochefort. "We work just enough. We don't want a stressful life," Father Antoine, a brewer at the abbey, once told Michael Jackson. "We try to achieve a balance, to have a sense of family." (Rochefort)

# Luxembourg

Bordered by Belgium, France, and Germany—three of the most important brewing countries in Western Europe, each of which represents a distinctive brewing tradition—little Luxembourg is uniquely situated to take advantage of several important traditions, yet its taste in beer runs generally to "pils." The largest brewing company was formed in 1975 as a merger of Brasserie Funck-Bricher of Luxembourg City and Brasserie Bofferding in the city of Bascharage, which were founded in 1764 and 1842, respectively. The merged entity was known as Brasserie Nationale in the twentieth century but has gone back to using the name La Brasserie Bofferding, which is the name of the flagship product, the 4.8 percent ABV Bofferding Lager Pils. It was decided at the time of the merger to close the Funck-Bricher brewery and focus the new company's entire brewing effort on this single brand and brewery. A seasonal Bofferding Christmas lager has been added in recent years.

Luxembourg's second largest brewing company is Brasserie de Luxembourg Mousel-Diekirch, the result of a 2000 merger of Brasserie Diekirch, based in the city of the same name, and Mousel et Clausen of Luxembourg City. The latter was itself formed by the 1747 merger of Brasserie Altmunster, founded in 1511, and Brasserie Mansfeld, founded in 1563. In 2002, Mousel-

Diekirch was acquired by AB InBev. The beers, all lagers, brewed at the Mousel brewery on the banks of the Alzette River in Luxembourg City, include Mousel Gezwieckelte Bier, Mousel Premium Pils, and Diekirch Premium, all rated at 4.8 percent ABV, as well as Diekirch Grand Cru Ambrée and Diekirch Christmas, both rated at 5.1 percent.

Other breweries in Luxembourg include Brasserie Simon in Wiltz, Brasserie Battin in Esch-sur-Alzette, and Brasserie Artisanale de Redange in Redange, which closed in 1880 but reopened in 2001. A newer microbrewery is Brasserie Cornelyshaff a Heinescheid, opened in 2003 in Kalborn. Simon products include 5.4 percent Okult No.1 Blanche witbier, 6.5 percent Simon Noël, 5 percent Waissen Durdaller hefeweizen, and 6.5 percent Wëllen Ourdaller dark ale. Battin's beers include 5.5 percent Battin Christmas and 5.2 percent Battin Gambrinus.

Of special interest to the beer traveler is Brasserie Salaisons Hotel "Beierhaascht," where one will find "located under the same roof, a brewery, an original restaurant, a traditional butcher's shop and a hotel built according to ecological guidelines." It is located in Bascharage, coincidentally the home town of Bofferding.

Above: The Brasserie Bofferding team celebrates the official launch of a new 24-bottle returnable case in 2012. (Bofferding)

Below: Bofferding releases its seasonal Christmas beers in October, waxing eloquently that "the days shorten. . . freshness is felt . . . Well yes the winter begins to point the tip of his nose! Nothing like the arrival of beers Noël Bofferding and Battin captivating and warm to warm up! Christmas beers lend themselves perfectly to the friendly atmosphere of the holiday of the year. Our master brewer has once again made a learned brewing that will delight the taste buds of beer drinkers." (Bofferding)

# France

In France people drink more wine than beer, but they drink far more beer per capita than the English, Germans, or Americans drink wine. Paris has an important brewing tradition that dates back to the Roman era, when Paris was called Lutecia. Gastronomy is an important, one might even say a defining, element in French culture. In November 2010, French gastronomy was added by the UNESCO to its catalog of the world's "intangible cultural heritage." One might also say that wine is a defining element of French gourmet cuisine. As the craft brew movement was sweeping the English-speaking world in the late twentieth century, few inroads were made in France

Though the history of brewing in traditionally wine-oriented France took a somewhat different path than it did elsewhere, that history is long and varied. By the time of the French Revolution, 28 breweries existed in Paris, principally in the marshy area called La Glacière (The Glacier), where they took advantage of the special qualities of the water and from which they harvested ice during the winter to store in caves for late summer use.

As with the production of wine, the brewing of beer in France is defined by regions. While brewing takes place throughout France, with brewpubs in Paris itself, there are two primary "appellations." These are, first, the Alsace, whose center is Strasbourg, and second, the Nord and Pas-de-Calais area in the northeast, adjacent to Belgium, centered around Lille and Armentieres.

In terms of total production and influence on the global export market, the Alsace is the most important brewing center in France. The Alsace is a place where German and French culture have been interwoven for centuries, with cultural roots that predate the formation of either country in its present form. The Alsace was part of Germany from 1871 to 1918 and from 1940 to 1944. Today, the cuisine of the Alsace represents the best of both cultures and is a

**Below: Enjoying a glass of Kronenbourg at an Alsace tavern.** (Carlsberg)

gastronomical wonderland. The heart of the Alsace is the city of Strasbourg, which is situated on the Rhine River across from Germany. The home of the European Parliament, Strasbourg is considered to be a political center of Europe.

Strasbourg is also home to Brasseries Kronenbourg, which is by far the largest brewing company in the France. Kronenbourg was founded in Strasbourg in June 1664, within a few yards of the old customs house on the banks of the Ill River, when Jerome Hatt, a newly certified master brewer, had just married and attached his seal to the first barrel of beer produced in his brewery, known as Brasserie du Canon.

By 1670, Hatt's brewpub had become the favorite meeting place of the people of Strasbourg. As the years rolled by, the Hatt family passed control of the brewery from father to son, and in 1850 Guillaume Hatt decided to move his brewing activities to the heights of the suburb of Kronenbourg.

By 1852, the beers of Alsace began to reach the cafés of Paris in significant quantities. In 1867, Kronenbourg was awarded several gold medals at the Universal Exhibition in Paris, which put them in the same class as competitors in Vienna and Munich.

Hatt's descendants went on to be active in the Alsatian brewing scene for the next three centuries. A member of the Hatt family remained at the helm of Kronenbourg until 1977, and Jean Hatt is recalled as founding another important Alsatian institution, Brasserie de l'Esperance in Strasbourg, in 1746.

In 1970, Kronenbourg became part of the Paris-based French food-products company Groupe BSN (Boussois-Souchon-Neuvesel), famous for Dannon and Danone Yogurt, which has been known as Groupe Danone since 1994. For many years, BSN owned not only

**Above: The Frog and Rosbif on Rue St. Denis in Paris is typical of a new generation of brewpubs opening across France. (Bill Yenne photo)**

**Kronenbourg is France's most internationally recognized brand. (Carlsberg)**

**Above: A glass of Pelforth Amber.**
**(Carlsberg)**

**Below: The Brasserie Duyck portfo-
lio includes Jenlain 6 with 6 percent
ABV, 6.6 percent Blonde d'Abbaye,
7.5 percent Blonde, 8 percent Or,
7.5 percent Ambrée, and 6.8 percent
Noël. All but Jenlain 6 are Bières de
Garde. (Duyck)**

Kronenbourg in France but also
Peroni in Italy, San Miguel in Spain,
and Alken-Maes in Belgium. In
2000, Kronenbourg was sold to
United Kingdom brewer Scottish
& Newcastle, which in turn was
purchased and split up by Heineken
and Carlsberg, with the latter
acquiring Kronenbourg.

At nearby Hochfelden, brewing had begun by at least AD 870,
and it was here in 1640 that Jean
Klein built his first brewery, an
establishment that was to remain
in his family for two centuries
until it was sold to the Metzger
family, who became linked by marriage to the Haag brewing family

from Inguiller in 1898. In 1927,
Louis Haag created the Meteor
brand, which today is the brewery's flagship label. The brands
include Meteor de Noël, Meteor
Pils, Meteor Lager, Grand Malt,
Wendelinus, Muse, and a wheat
beer called Blanche de Meteor.

Through the twentieth century, the second biggest brewing
operation in Alsace was centered
on the Brasserie Fischer. Founded
in 1821 in Strasbourg, it moved
to Schiltigheim in 1854. Acquiring
the Grande Brasserie Alsacienne
d'Adelschoffen in 1922, the company was renamed as Groupe
Pêcheur, trading the German
"Fischer" for its French transla-
tion. The group was absorbed into
Heineken in 1996, but
the leading brands were
retained. These include
the flagship Fischer
Tradition Amber and
Blonde, sold in swing-
top bottles, as well
as Fischer LaBelle,
Fischer Bière de
Noël Christmas
beer, the tequila-fla-
vored Desperados,
and Adelscot, aged
in Scotch whis-
key barrels under
the Adelschoffen
name. In 2009,
Heineken closed
the original
Brasserie
Fischer

plant, concentrating all of its Alsatian production at the Brasserie de l'Esperance facility.

Another major Heineken presence in France takes us north, into the Département du Nord, to the Brasserie Pelforth in Mons-en-Baroeul, a suburb of the city of Lille. When founded in 1914, it was called the Brasserie du Pelican, after the "dance of the Pelican," which was a contemporary dance craze. A pelican still graces the label. During the coming years, the company grew into one of France's largest brewers through acquisitions of smaller breweries, such as Brasserie Carlier of Coudekerque-Branche, and expanding distribution. Pelforth, created as an English-style ale, was first brewed in 1935.

Having been renamed Brasseries Pelforth in 1972, the company was itself acquired, going through a series of parent companies, beginning in 1980 with Groupe Brasseries et Glacières Internationales (BGI), a French colonial brewing conglomerate. Originally Brasseries et Glacèries d'Indochine, BGI was famous for its 33 Export lager, widely available in Indochina and throughout Southeast Asia. Pelforth was acquired by Heineken in 1986, and BGI was bought in 1990 by the Castel Group, the largest specialty wine retailer in France and a company with an increasing interest in brewery acquisitions. The brewery at Mons-en-Baroeul

still produces Pelforth Blonde and Pelforth Brune as well as 33 Export. A Pelforth Amber was introduced in 2003.

As it does with breweries that it owns around the world, Heineken also brews its own signature lager at its breweries in France.

With the discovery of beer culture as an essential element of French gastronomy, the beer traveler will wish to seek out the smaller independent brewers which have been flourishing in France in the twenty-first century. To do so, one need not leave the Département du Nord—indeed, this is the appellation d'origine contrôlée for one of France's most unique beer styles.

Bière de Garde, which literally means beer to be squirreled away, is also called farmhouse beer because it was traditionally made at home. Even after it was made commercially, it was made on a very small scale. It is a golden unfiltered ale made with top-fermenting yeast, and may be bottle conditioned for

**Above: The Brasserie Duyck in Jenlain. (Jenlain)**

**Below: One for the largest brewing companies in the Alsace for much of the twentieth century, Brasserie Fischer in Schiltigheim, was acquired by Heineken in 1996. Though the brewery was later closed, the brand and the signature flip-top bottle, seen here in an early 1990s image, remain. (Collection of the author)**

Above: The Brasserie Duyck bottle line. (Duyck)

Below: The Brasserie Thiriez portfolio includes 6.5 percent ABV Blonde d'Esquelbecq, 8.5 percent Dalva, 5.5 percent Etoile de Nord, 5.5 percent Les Québecoises, and 5.5 percent La Rouge Flamande. D'Esquelbecq and Etoile de Nord are farmhouse-style ales. (Thiriez)

a long period of time. It is similar to the beer style that the Wallonian tradition in nearby Belgium calls saison. Bière de Garde is typically in the 7 to 9 ABV range. Notable bières de garde include Trois Monts from Brasserie de Saint-Sylvestre in the town of Saint-Sylvestre-Cappel and Jenlain from Brasseurs Duyck in Jenlain.

Though recently "discovered" by a new generation of gastronomes and beer lovers, Brasseurs Duyck has a long history of bières de garde. It was founded in 1922 by Felix Duyck, the son of Leon Duyck, who started brewing in the town of Zegerscappel in Flanders in 1900. Felix established his farm and brewery in Jenlain, where he developed and brewed a bière de garde. Bottling, in discarded champagne bottles, began

in 1950. Felix's son Robert took over in 1960 and met increased demand with a modest expansion of his father's original brewery. Beginning in 2002, fourth-generation brewer Raymond Duyck began a substantial expansion, adding a dozen new fermentation tanks and expanding distribution across northern France. There are now dozens of establishments from Abbeville and Amiens, and from Lille to Fontainebleau, serving Jenlain beers on tap. In Jenlain itself, visit La Jenlinoise on Place de Jenlain. In Paris, one might even stop in at the Estaminet Jenlain at 119 rue de Reuilly.

Chronologically part of the global craft brewing movement, Brasserie Thiriez was founded in 1996 by home brewer Daniel Thiriez, although he started his

business in a classic farmhouse brewery which originated immediately after World War II and hosted several companies through the years. It is situated in Esquelbecq in the Département du Nord, a short distance from Dunkirk and the Belgian border. In 2006, Thiriez added a 20-hectoliter brewhouse and installed four new 40-hectoliter fermenters.

Since the 1990s, French cities, especially in the north, have seen a number of brewpubs come and go and flourish. Notable for the beer traveler are a couple of chains of brewpubs. The Frog Pubs group originated in 1993 as "an MBA project researching the potential for an English pub & micro-brewery in central Paris." Beginning with the original Frog & Rosbif at 116 Rue Saint-Denis in Paris, the chain expanded to additional locations in the Paris area as well as Bordeaux and Toulouse.

Another chain, similar in concept, but larger in scope, are the Trois Brasseurs ("Three Brewers") brewpubs. From a first pub in Lille, they expanded to more than two dozen in metropolitan France, mainly in the north, but as far afield as the Loire Valley and Grenoble. There are Trois Brasseurs brewpubs in French overseas territories from Tahiti to New Caledonia and eight in Canada, mainly in Montreal.

French cuisine is the haute of haute cuisine, and although

French wine is inextricably linked with the French culinary arts in the mind of foreign epicures, beer is no longer something relegated to the rustic farmhouse. French beer has a well-deserved place at the exalted white linen–covered table. As with French cuisine and French viticulture, French beers are brewed with care and attention to the fine subtleties of flavor. Like French wines, French craft beers are carefully designed to accompany specific foods, and they succeed superbly. As Michael Jackson so aptly pointed out, the word "brasserie," which implies a typical Parisian café, actually means "brewery." What do Parisians typically drink in their brasseries? Bière.

Above: Two beers from Les Brasseurs de Gayant in Douai, are 4.5 percent ABV Amadeus, a bière blanche, and La Goudale, a Bière de Garde. (US Brands)

Below: Service with a smile at Brasserie Thiriez. (Thiriez)

# Southern Europe

From the time of the great Greek, Etruscan, and Roman civilizations, the lands surrounding the Mediterranean were a region of vineyards, where wine and winemaking flourished and became an art form. Italy, which the ancient Greeks called Oenotria (the land of wine), now produces more wine than France. Spain, Portugal, and Greece are also major producers and consumers of wine, with Spain having more acres of vineyards than either France or Italy. In fact, in Portugal 15 percent of the population is involved in the wine industry.

Since the 1970s, however, beer has been taking away market share from wine, especially in the cities of Spain and Italy where beer drinking developed a cachet among younger drinkers that will remind the beer traveler of the brasserie scene in Paris. Annual Spanish beer consumption increased from 32 liters per capita in 1970 to 51 in 1978 and stands at around 70 today. In Italy, annual consumption of beer increased from 14 to 24 liters per capita between 1975 and 1985, and stands at around 30 today.

Between 1980 and 2010, annual beer production (mainly of pale lager) increased from 19 million to 35 million hectoliters in Spain, from 7 million to 13 million hectoliters in Italy, from 2 million to 8 million hectoliters in Portugal, and from 1.7 million to 4 million hectoliters in Greece.

Meanwhile, and certainly of special interest to the beer traveler, Zak Avery, writing in *The Beer*

Below: Workers at the Menabrea brewery Menabrea brewery in Biella in Italy's Piedmont pose for a group photo in November 1897. (Menabrea)

*Connoisseur* magazine, estimates that the number of craft breweries in Italy alone increased tenfold from around 40 in 2005 to 400 by 2010. Even with industry consolidation eating away at this number, the trend is remarkably good news for a beer traveler in a land where craft brewing was virtually unknown at the end of the twentieth century.

In Spain, indigenous brewers had an opportunity to grow without competition until the second half of the twentieth century. The Spanish market remained closed to imported beer until 1960, with the exception of small quantities of German and Danish beer sold there in the 1950s. In 1960, Spain signed a trade agreement with the Benelux nations, and a small volume of Heineken entered the market during the next decade. In the early 1990s, Spain eclipsed France and the components of the former Czechoslovakia as Europe's fourth-largest brewing nation after Germany, Russia, and the United Kingdom. Since 2010, it has slipped into a tie with Poland for this posi-

tion. Spain closed out the twentieth century with around 26 million hectoliters in annual production and topped 35 million a decade later.

Spain's largest brewing companies are Sociedad Anónima Damm (SA Damm), and Heineken-owned Cruzcampo. The former was founded in Barcelona in 1876 by Alsatians August Kuentzmann Damm and Joseph Damm. The Damm family later joined with other brewers, such as Juan Musolas (brewers of La Bohemia) and Cammany y Cia (who later came to be known as Puigjaner y Cia). Having evolved from this merger, today's SA Damm group offers the market a wide range of products that have allowed it to establish a reputation

**Above: The stylish packaging of the 4.8 percent ABV A.K. Damm lager, brewed by Damm in Barcelona. (Damm)**

**Above: An icy bucket of Peroni Nastro Azzurro, one of Italy's leading lagers. (Jason Alden/One Red Eye)**

**Below: The lager portfolio of Cervezas Alhambra in Granada, Spain includes 5 percent ABV Especial, 5.4 percent Premium, 5.4 percent Negra, 7.2 percent Mezquita, and non-alcoholic Sin. (Alhambra)**

as one of the leading brewers on the Iberian peninsula. Its breweries in Barcelona, Santa Coloma de Gramenet, and El Prat de Liobregat are well equipped.

Damm's flagship lager is Estrella Damm, formerly known as Estrella Dorada, which dates back to 1876. Marketed in variants that range between 4.6 and 5.4 percent ABV, it is a product that the beer traveler will notice to be one of the most popular beers in the Barcelona area, its bottles and cans displaying a characteristic golden star. At 7.2 percent ABV, Voll-Damm imitates a German märzen. Xibeca is a 4.6 percent table beer often seen in large one-liter bottles. Xibeca was created to be consumed with meals as a cheap substitute for red table wine when table wine

prices rose at the end of the 1960s. It became very popular in Catalonia during the 1970s. Other Damm products include 4.6 percent ABV A.K. Damm, 5.4 percent Bock Damm, and 4.8 percent Weiss Damm, a hefeweizen.

Cruzcampo was founded in 1904 by Roberto Osborne and Agustín Osborne, taking its name from the Cross of the Field (Cruz del Campo), which still stands in the field next to the brewery in Seville. As Damm has an association with Catalonia, Cruzcampo is strongly associated with Spain's Andalusian region. With breweries in Arano, Jaén, Madrid, Seville, and Valencia, the brands from the company, all lagers, include 2.4 percent ABV Cruzcampo Light, 4.8 percent Cruzcampo Pilsener, 5 percent

Cruzcampo Lager, and 6.4 percent Cruzcampo Gran Reserva.

Madrid's big brewing company is Grupo Mahou-San Miguel, which traces its roots back along the lineage of two companies. In 1953, Andrés Soriano, the head of the San Miguel Brewery in Manila, the founder of Philippine Airlines, and an associate of Spain's Generalissimo Francisco Franco, licensed the use of the San Miguel brand name to a newly formed brewing company in Lerida, Spain, called La Segarra. This company became San Miguel, Fábricas de Cerveza y Malta in 1957 and opened a second brewery in Malaga in 1966. In 1992, France's Groupe Danone began the acquisition of San Miguel.

Meanwhile, back in 1890, Hijos de Casimiro Mahou (Sons of Casimiro Mahou) was established in Madrid by a French brewer from Lorraine and named for his three children, Enrique, Luis, and Carolina. Enrique took over in 1943, and a new brewery on Paseo Imperial in Madrid was opened in 1962. By 2000, Mahou owned 30 percent of San Miguel, and proceeded to acquire the 70 percent owned by Group Danone.

Grupo Mahou-San Miguel brands, all lagers, are still marketed under their earlier brand names although both breweries produce a San Miguel Especial, rated between 4.2 and 4.4 percent ABV. The Mahou portfolio includes 5.5

percent ABV Cinco Estrellas, 4.8 percent Clasica, 5.5 percent Negra (dark lager), and 1 percent Sin. From San Miguel there was 4.4 percent San Miguel Fresca and 6.2 percent San Miguel Selecta.

In Portugal, there are two brewing companies sharing 90 percent of the market. They were formed by a consolidation of many smaller breweries that was engineered by the government after the industry was nationalized following the 1974 coup. Both privatized in the 1990s. Uniao Cervejeira (Unicer) remains independent, while Sociedade Central de Cervejas (SCC) was acquired by Scottish & Newcastle in 2003, which was in turn acquired by Heineken, making SCC a Heineken company since 2008. Vialonga-based SCC was founded in 1934, while Unicer, based in Sao Mamede de Infesta, dates back to Companhia Uniao Fabril, formed by an earlier consolidation in 1890.

The Unicer portfolio, all lagers, includes 5.1 percent ABV Cristal,, 4.2 percent Cristal Preta (Dark), and 6.8 percent Super Bock Abadia Gold. The SCC flagship brand is Sagres, named for the village in southwestern Portugal where the great navigators of the fifteenth and sixteenth centuries were trained. The brand was created for the 1940 Portuguese World Exhibition, and was widely exported to

These pages: From the Damm portfolio, 3.5 percent ABV Saaz, and 5.4 percent Bock. Variations on the flagship Estrella Damm range from 4.6 to 5.4 percent. (Damm)

Above and right: The classic "Moretti Man" is based on a snapshot of an anonymous gentleman taken (with permission) by Manazzi Moretti in 1942. In recent years, he has been redrawn to be more erect, and no longer to be seen blowing at the head atop his beer. (Heineken Italia)

Below: Birra Moretti and Birra Ichnusa of Sardinia, two selections from the Heineken Italia folio, which also includes Dreher. (Heineken Italia)

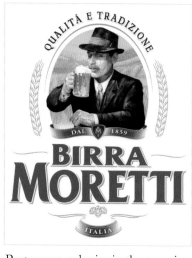

Portuguese colonies in the twentieth century. Specific beers, all lagers, include 5.1 percent ABV Sagres Cerveja and 4.4 percent Sagres Cerveja Preta (Dark) as well as 6.2 percent Bohemia and 4.5 percent Imperial Cerveja Viva.

Italy's largest brewing company is Birra Peroni Industriale, founded in Vigevano in 1846 by Francesco Peroni and relocated to Rome in 1867 by Giovanni Peroni. In 1924, Peroni built a new plant in Bari, and in 1929 took over Meridionali in Naples. The growth of Peroni continued with its acquisitions of the Dormisch Brewery in the northern Italian town of Udine, the Itala Pilsen Brewery of Padova, the Raffo Brewery in Taranto, and the Faramia Brewery in Savigliano. In the 1960s, Peroni opened three new state-of-the-art breweries in Bari, Rome, and Padova.

From the 1950s to the present day, Peroni's flagship lager, 4.7 percent ABV Birra Peroni, has been the most widely recognized brand in Italy, followed by Peroni Nastro Azzurro, a 5.5 percent ABV lager introduced in 1963 and named for the nastro azzurro, or blue ribbon, won by the Italian ocean liner *Rex* for crossing the Atlantic at record speed in 1933.

During the late twentieth century, Peroni was part of a Big Four of close competitors for the Italian market that also included Dreher, Wührer, and Moretti. While Peroni controlled a quarter of the market, Dreher held nearly a fifth. The company was founded in Trieste in 1896 by Anton Dreher, the great Vienna brewer who defined Vienna-style lager. At the time, Trieste, part of Italy since World War II, was the leading port of the Austrian Empire. Wührer, meanwhile, was founded in Brescia in 1829, and Moretti was started in 1859 by Luigi Moretti in the northern Italian city of Udine. The Moretti label's famous painting of the man with the mustache is based on a photograph taken in a café in Udine in 1942 by Manazzi Moretti. The image, though it has been altered a time or two, has long been familiar throughout Italy, Europe, and much of the world.

In a round of acquisitions over three decades, the Big Four gradually ceased to be independent. When Dreher was bought by Heineken in 1974, its outdated breweries in Turin and Trieste were closed, and Heineken and Dreher production was concentrated in the breweries in Pedavena, Massafra, Macomer, and Genoa. Meanwhile, Wührer was acquired by Peroni in 1988.

In 1989, a large stake in Moretti was acquired by Labatt, Canada's second largest brewer. Labatt also acquired Prinz Bräu in Crespellano, Italy. Labatt itself became part of Interbrew in 1995, and Moretti entered the Heineken portfolio in 1996.

Finally, in 2005, Peroni and its brands were acquired by SABMiller, and the bulk of the Italian beer market came to be foreign owned. Today, Birra Peroni and Nastro Azzurro continue to head the Peroni all-lager lineup, which also includes 3.5 percent ABV Leggera, 6.6 percent Gran Riserva, and 4.7 percent Wührer Lager.

Dreher, still brewed as a 4.7 percent ABV lager, is kept by Heineken in a separate portfolio from Moretti, being listed within Milan-based Heineken Italia. Other brands in this portfolio are Ichnusa and Messina, also both 4.7 percent lagers. The Heineken Moretti portfolio, brewed in Udine, includes the flagship 4.6 percent ABV Birra Moretti as well as 5.6 percent Moretti Sans Souci, both of which are pale lagers. Also included are Moretti La Rossa, a 7.2 percent

amber ale, and Moretti Grand Cru, a 6.8 percent Belgian-style ale.

Mention should also be made of smaller Italian brewers that are still going strong in the twenty-first century. An example is Birra Forst (Brauerei Forst) in the Italian province of Alto Adige, founded in 1857 by Johann Wallnöfer and Franz Tappeiner in Algund, when the area was still part of Austria. In 1863, the company passed to the entrepreneur Josef Fuchs, who built the current plant in Forst. Another example is the Menabrea brewery in Biella in the Piedmont region was established in 1846, acquired by Giuseppe Menabrea and Antonio Zimmermann in 1864, and recognized by Italian King Umberto I, who knighted Giuseppe in 1882. In 1896, the

Above: Birra Peroni, a 4.7 percent ABV lager has been the Rome-based company's flagship product inside Italy for decades. (Classic label from the author's collection)

Below: A dapper waiter serves up a cold bottle of Peroni's Nastro Azzurro. This 5.1 percent lager is widely available in Italy and in the export market. (SABMiller/ One Red Eye/Philip Meech)

Above: The brewery at Birra Peroni in Rome. (SABMiller)

Below: Machete, a 7.8 percent ABV double IPA from Birrificio del Ducato in Roncole Verdi di Busseto, Italy; and four from Menabrea: Birra di Natale, Anniversario Lager, Anniversario Amber, and Anniversario Strong. (Ducato/Menabrea)

company lineage passed to the brothers-in-law Emilio Thedy and Agostino Antoniotti, husbands of Eugenia and Albertina Menabrea, and the company remains in the family into the twenty-first century. Their beer is distributed by Forst. The Menabrea brand portfolio includes three Menabrea 150th Anniversario beers,

the 4.8 percent ABV Lager, and the 5 percent Amber.

The focus of interest in the Italian brewing scene since the turn of the twenty-first century, especially for the beer traveler, has turned to the arrival of the craft brewing movement, which first took root in northern Italy. The first brewpub in the area was Birrificio Italiano, which

began brewing in 1997 in Lurago Marinone, near Lake Como. Their flagship beer is a 5 percent ABV pilsner called Tipo Pils, while other products include a 6.4 percent fruit beer called Cassissona, and a 6 percent dunkelweizen called Vùdù.

Other notable craft breweries in the north are Birrificio Lambrate in Milan and Birra Busalla in Savignone, north of Genoa, which was founded by Giacomo Ricchini, Giovanni Bagnasco, and Sisto Poggi but closed from 1929 to 1999.

Perhaps the best-known microbreweries in the north are Teo Musso's several Baladin brewery restaurants. Beginning in Piozzo in the Piedmont region, about 40 miles south of Turin, they have proliferated to Turin, Milan, Rome, and even to New York City. Some random selections from Baladin's extensive list include 6.8 percent ABV Baladin Nora, 9 percent Noël Baladin, 8 percent Super Baladin, and a range of Xyauyù barleywines with ABV in the 12 to 14 percent range. Among these is the delicious 14 percent Xyauyù Fumé, aged in Islay Whisky barrels.

Birrificio del Ducato was founded in 2007 in Roncole Verdi, a small village near Parma that is the birthplace of the composer Giuseppe Verdi. As the founders describe themselves, Giovanni Campari is "the radical and visionary brewmaster . . . holding a BA in Food Science and Technology and

a past as home brewer." Manuel Piccoli is the brewery's "executive entrepreneurial mind with a precise vision for growth and development." In 2008, their 8.2 percent ABV Verdi Imperial Stout ranked first at the European Beer Star in Germany, "the first time in history an Italian artisan beer has ever won an international contest." Their Verdi Imperial Stout Black Jack weighs in at 10 percent ABV. In 2010, the 4.8 percent ABV Via Emilia won a silver medal at the World Beer Cup in the United States. In 2011, Ducato received the accolade for being Italy's Brewery of the Year for its second time. Other products include 13 percent ABV L'Ultima Luna barleywine, 5 percent Nuova Mattina (New Morning) saison, 4.8 percent Sally Brown stout, and 5.2 percent Sally Brown Caffe Baracco stout.

Moving south, the city of Rome has developed a world-class craft beer scene. In addition to Baladin, the beer traveler will want to visit Brasserie 4:20 and Revelation Cat, both founded by Alex Liberati. Birra Del Borgo, located in Borgorose, about 40 miles north of Rome, has a worthy presence in the cafés and beer bars in Rome. Founder Leonardo Di Vincenzo became a homebrewer while studying biochemistry in the university and started his own commercial brewery in 2005. In 2007, he collaborated with Manuele Colonna, owner of the

Above: The 8.2 percent ABV Verdi Imperial Stout from Birrificio del Ducato. (Ducato)

Below: Extra ReAle, a 6.2 percent IPA from Birra del Borgo in Borgorose, Italy. (Del Borgo)

**Above left: Mythos Hellenic Lager from Thessaloniki. (Mythos)**

**Above right and below: For more than a century before its demise in the 1970s, Fix Hellas was Greece's signature lager. Since 2010, with a new brewery at Ritsona (below), it has enjoyed a strong comeback. (Fix Hellas)**

Macchesietevenutiafa pub, to open the Bir e Fud pizzeria in Rome's Trastevere district. As Di Vincenzo describes it, this was "the first venue in Rome to bring customers a different way to drink beer along with quality food. . . . [W]e wanted to prove that the 'pizza and beer' pairing, a classic in Italy, could find a new meaning by matching craft beers with gourmet pizzas and other simple but well prepared dishes."

In 2009, he collaborated with Teo Musso of Baladin and chef Gabriele Bonci to start Open Baladin Roma near Rome's Campo de Fiori. In 2011, Di Vincenzo teamed up with Teo Musso and Sam Calagione of Dogfish Head to launch La Birreria, a rooftop brewpub in New York. The extensive Birra Del Borgo line

includes three names which Di Vincenzo calls his "most famous recipes." These are the Duchessa saison family and DucAle, an 8 percent ABV Belgian dark ale, as well as the ReAle family of pale ales and IPAs. Duchessa herself comes in at 5.8 percent, while Duchessic Ale is 5.9 percent. The ReAle family ranges in ABV from 6.2 percent for ReAle Extra to 6.5 percent for the last few Anniversario ReAles.

In Greece and the surrounding islands, the iconic brand for more than a century was Fix Hellas (called simply Fix), a lager first produced in Athens in 1864 at the brewery of the Bavarian expat Johann Karl Fuchs (Karolos Ioannou Fix in Greek). Beginning to lose market share to the imports in the 1970s, the Fix Hellas brewery folded in 1983. The landmark brewery was demolished in 2002 after a losing battle to preserve it. In the meantime, the Fix Hellas brand

was briefly revived in 1995 by the Kourtakis winery interests under the name Olympiaki Zythopoiia (Olympic Brewery). In 2009, three Greek businessmen, Ilias and George Grekis and Yiannis Chitos, took over the languishing company and built a new state-of-the-art brewery in Ritsona. The product line, all lagers, includes the 5 percent ABV flagship brand, Fix Hellas, as well as 5.2 percent Fix Dark, 4.5 percent Proton, and 5.2 percent Z.

Fix was soon restored to its position as the leading brand in Greece, ahead of the Mythos brand. The latter, 4.7 percent ABV Mythos Hellenic Lager Beer, was launched in 1997 and is brewed in Thessaloniki in northern Greece at a brewery started by Henninger in 1970. The brand and brewery have been owned by Carlsberg since 2008.

As Eleni Dessipri of Mythos explains, "Mythos beer was launched in 1997 in Greece and due to its excellent product quality and its genuine, unconventional brand personality, Mythos is now one of the top three beer brands in the Greek Market. Mythos is a refreshing, easy to drink lager beer, characterized by a fruity aroma, bright golden color and rich head, and due to its high quality and unique taste Mythos has been awarded in international beer competitions."

There is also a small craft brewing segment in the Greek market. Some examples of Greek microbreweries are Craft, at several Athens locations; Piraiki in nearby Piraeus; Hillas in Rodopi; and the Rethymnian Brewery, near Armeni on the island of Crete.

**Above: The portfolio of Piraiki in Piraeus includes 5 percent ABV pils, pale ale and lager. Having been certified as organic , the beer is referred to on the label as "Bio Beer." (Piraiki)**

**Below: Enjoying a round of Mythos Hellenic Lagers at the beach. (Mythos)**

# The Middle East and North Africa

**Below: Located in Cumra, the largest malting facility in Turkey is owned by Anadolu Efes, the holding company of the Efes Beverage Group. It opened in 2007 with a capacity of nearly 1.8 million bushels, and was expanded to 2.6 million bushels, opening in October 2008, just ahead of the barley harvest. (Efes)**

According to archeologists, the story of brewing in the Middle East begins in ancient Persia and Sumeria at least 7,000 years ago. Often cited by beer historians is a Sumerian tablet from the second century BC containing a poem to Ninkasi, the goddess of brewing. This document also has the oldest-known beer recipe, which was produced for the first time in centuries by Fritz Maytag at Anchor Brewing in San Francisco in 1989.

Fast forwarding to the nineteenth century, we find that most of the major commercial breweries across the region were founded and operated by European colonial powers, although in Egypt an indigenous style called boosa was being brewed by the Bedouins. German and British breweries were opened in Egypt and elsewhere, just as the French were developing the sunny hills of Morocco, Algeria, and Tunisia for viticulture. Many of the European breweries in the region became locally owned and operated by the mid-twentieth century, although Heineken still has a prominent presence.

In Egypt, there were two notable European breweries in the twentieth century, Bières Bomonti & Pyramides in Cairo and the Crown Brewery which was chartered in Belgium and began brewing in Alexandria in 1897. In the early twentieth century, they jointly marketed a mass-market Pilsen-style lager known as Stella (Star). During World War II, while brew-

eries in Europe suffered or closed, production of Stella increased sixfold because of the huge number of Allied troops stationed in Egypt. Heineken, which had held a share of Bomonti & Pyramides since before the war, continued to play a role until the company was placed under state control in 1956.

Crown was nationalized in 1963, with the name changed to Al Ahram Beverages in 1985. Privatized in 1997, the company was acquired by Heineken in 2002. The largest brewing company in Egypt, Al Ahram now operates a brewery in El Obour and non-alcoholic beverage plants in Badr and Sharkia. The brewery's flagship brand continues to be Stella, a 4.5 percent ABV lager. Also produced are 4 percent ABV Sakara Gold and 10 percent ABV Sakara King. Non-alcoholic brands include Amstel Zero and Fayrouz.

Egypt's second-largest brewery, albeit much smaller, is the Egyptian International Beverage Company, brewers

of the Luxor brand, which was founded in 2005.

Heineken also owns the Lebanese Brasserie Almaza, founded in 1933 in Bauchrieh by the Jabre family as Brasserie Franco-Libano-Syrienne. The brewery produces the widely-exported Almaza Pilsener, a 4.2 percent ABV lager, and the 6 percent ABV Almaza Pure Malt.

By the end of the twentieth century, Turkey was the largest brewing nation in the Middle East, with 4.8 million hectoliters in annual production, followed by Egypt and Israel, with annual outputs below a million hectoliters. Iraq was also once a major

**Above: Nadim Khoury studied brewing at the University of California in Davis and founded the Taybeh Brewery in the West Bank village of the same name in 1994. In 2005, he launched his annual Oktoberfest, which has become a popular event in the Palestinian Territories. (Taybeh)**

Above: Efes is the ubiquitous brand in Turkey, and the company claims it to be the "No.1 Mediterranean Beer." (Efes)

Below: The beer of ancient Egypt as transcribed from the hieroglyphics, circa 1922, by H.F. Lutz. On the left, the top line translates as the Egyptian word for "brewing beer." The center line contains three words for "brewer," and the bottom line is "those who crush [mash] the grain for beer." On the right are names for three beer styles. From the top, they are translated as "beers used exclusively for religious purposes;" "beer which does not sour," or "beer of eternity:" and "beer of the goddess Maat." (Author's renderings from H.F. Lutz drawings)

brewing country. With the rise of fundamentalist Islam, particularly in the twenty-first century, many breweries across North Africa and the Middle East have been forced to close or maintain a low profile.

Turkey's leading brewing company, with an 80 percent market share, is the Efes Beverage Group, named for the ancient city of Ephesus, near Izmir, where it operates one of its breweries. The first Efes brewery, Erciyas Biracilik ve Malt Sanayii, was established in Istanbul in 1966 and began marketing its beers in 1969. The company also maintains a maltery at Afyon and a hop facility at Tarbes. The flagship Efes Pilsen, a 5 percent ABV lager that is available not only in Turkey but is exported throughout the Middle East, Africa, and Europe. The company also brews Bomonti and Marmara lagers and Gusta wheat beer as well as various European brands under license. Efes also operates in Kazakhstan, brewing

the Karagandinskoye brand at its brewery in Karaganda.

The second-largest brewing company in Turkey is Turk Tuborg in Izmir, formerly owned by the Danish Carlsberg Group, but now owned by the Central Bottling Company of Israel, which also owns Israel's second-largest brewer. Originally a soft drink bottler, Central Bottling became a brewer through the acquisition of Israel Beer Breweries Limited, which began operations in 1992 at Ashkelon in Israel, and which was partially owned by Carlsberg until 2008.

An example of a locally established brewing company is Tekel Birasi, which was founded in Turkey in 1890 and was government owned until 2004.

The largest brewing company in Israel is Tempo Beer Industries, whose flagship brand is Goldstar, a 4.9 percent ABV lager that is certified as kosher. Goldstar is described by the company as

"Israel's most beloved and best selling beer." Since 1968, Tempo has also brewed Maccabee, a 5 percent ABV lager which it calls "Israel's official alternative beer." As Tempo is Heineken-owned, they also import that company's leading products as well as those of other international brewers.

Since craft brewing arrived in Israel, the texture of the beer drinking experience has changed. Among the microbreweries are the Golan Brewery, founded in 2006 at Qatzrin in the Golan Heights, and the Lone Tree Brewery, established at Kfar Etzion, 15 minutes from Jerusalem, in 2010. Golan founders Naftali Pinchevsky and Haim Ohayon note that there is archeological evidence of brewing in the Golan Heights in ancient times, and they see themselves as carrying on a tradition.

Named for an ancient oak that is the symbol of Gush Etzion, Lone Tree was founded by Scottish expats Myriam and David Shire and American immi-

grants Susan and Yochanan Levin. The Scottish and American roots of the founders can be seen in the product line, which includes Extra Oatmeal Stout, California Steam Ale, English Northern Ale, Belgian Piraat Ale, India Pale Ale, and London Pale Ale.

Meanwhile, very near geographically to the Israeli microbreweries in the Palestinian Territories is the Taybeh Brewery, which began brewing

Above: Classic Middle East labels include lagers from Efes and Turk Tuborg, as well as a non-alcoholic beer from the Arab Company of Jordan. (Author's collection)

Below: A range of products from the Lone Tree Brewery in Kfar Etzion, Israel include Extra Oatmeal Stout, California Steam Ale, English Northern Ale, Belgian Piraat Ale, India Pale Ale, and London Pale Ale. (Lone Tree)

**This page: Nadim Khoury's annual West Bank Oktoberfest. (Taybeh)**

in 1995 in the West Bank village of Taybeh, about 20 miles north of Jerusalem. It was founded by Nadim Khoury, a home brewer who had studied brewing at the University of California in Davis, where many American brewers were trained in the late twentieth century. Taybeh lagers include Golden, Light, Amber, Dark, and a non-alcoholic beer introduced in 2007 for the local Palestinian Muslim market. Since 2005, Khoury has produced an annual beer festival based on the Munich Oktoberfest.

At the opposite side of North Africa is Morocco's Brasseries du Maroc in Fes, Tangier, and Casablanca, which market their Spéciale Flag, Stork, and Casablanca lagers as well as producing Heineken under license.

Until recently, the beer traveler might have found various other locally produced lagers in other countries in the Middle East. In Syria, these would have included Aleppo-brewed Al-Shark and Barada from Damascus. In Jordan, Amstel has been brewed since 1958 by the Jordan Brewery Company, founded in 1955 by Abujaber & Sons, a Jordanian group at Zerka north of Amman.

In Iran, where consumption of alcoholic beverages is as widespread as it is illegal, there are several well-established producers of non-alcoholic beer, including Arpanoosh, brewers of the Istak brand, and Danjeh Aria, who brew the Hoffenberg brand. Iran's largest non-alcoholic brewing company, Behnoush, brewed real beer before 1979 and can be expected to resume when or if Prohibition is relaxed.

Across the former Soviet Central Asia, a number of state-owned breweries were privatized in the 1990s, and many are still active. In Kazakhstan, the beer traveler seeking local beer might enjoy the products of the Shymkentpivo brewery in Shymkentskoye, the Efes brewery in Karagandinskoye, or of the Carlsberg-owned brewery in Derbes.

In 1997, the French beverage conglomerate Groupe Castel acquired and began investing heavily in breweries in Abovyan,

Armenia, and suburban Baku in Azerbaijan with an eye toward spinning them off. The former, brewers of Kotayk, Armenia's biggest brand, was sold to the wealthy Armenian entrepreneur Gagik "Dodi Gago" Tsarukyan in 2011. The latter was sold to St. Petersburg-based Baltika in 2008. Beers produced here include Xirdalan Gara (Black), Xirdalan 7/7, 33 Export, Bizim Piva (Our Beer), and Afsana (Legend).

**Above: Belgian Piraat Ale from Israel's Lone Tree. (Lone Tree)**

**Above left: The classic label from Maccabee, Israel's leading brand. (Tempo Beer Industries)**

**Below: Baltika began brewing at this facility in Baku, Azerbaijan in 2009. (Baltika-Baku)**

# Africa

To use an obvious metaphor, the beer traveler's tour of Africa is likely to be like that of the proverbial blind men assessing the proverbial elephant. The shape and texture of the African beer scene is likely to be quite different wherever on the continent one sets down. Because Africa, unlike Europe and North America, is not conducive to a transcontinental road trip—or transcontinental product distribution—one experiences much more of a regional character. Having said that, the pale, golden beer style generally referred to as international lager certainly predominates in each region, with each of these having its favorite brands.

As with the blind men and their elephant, Africa is actually four continents, one for each point of the compass. North Africa, distinctly different from the rest of Africa, has more in common with the Middle East and is covered in the previous chapter. Southern Africa is economically dominated by the Republic of South Africa, the largest economy on the continent. In West Africa, the economic center is Nigeria, the second-largest economy in Sub-Saharan Africa and the most populous African country. Meanwhile, Sub-Saharan East Africa is economically centered in Kenya.

As was the case throughout the world, early African brewers used available materials in their making of beer, and today Sub-Saharan commercial brewers continue to

Below: Nile Special has been the flagship brand of Nile Breweries in Uganda for over 50 years. It is an eight times Gold Award and double Grand Gold winner at the Brussels-based Monde Selection International. (SABMiller/David Parry/One Red Eye)

use ingredients which are compatible with the climate and the crops being grown. For example, maize and sorghum malt is widely used instead of barley malt, even in the operations of European-based brewers on the continent. Beer made using cassava as a fermentable starch, used by African home brewers for generations, has been introduced by large commercial breweries from Ghana to Mozambique.

While pale lager does predominate as a beer style, dark beer is also popular, especially Guinness Foreign Extra Stout, which is brewed in Nigeria and elsewhere. Other indigenous stouts can also be found as well as various dark lagers which are known generically in Africa as "opaque beer." In southern Africa, one often encounters chibuku, also called scud or shake shake, which is a traditional opaque beer made from sorghum and maize that is often sold in paper cartons.

In Rwanda, an adventurous beer traveler might seek out and sample Urwagwa, a beer made from fermented banana juice and sorghum flour. In Uganda and elsewhere in East Africa, one might find Waragi, a triple-distilled, millet malt–based liquor (40 percent ABV). Just ask for "the Spirit of Uganda."

In so vast a region, it is hard to generalize about the overall nature

**Above: Laurentina has been a favorite lager in Mozambique since it was introduced in 1932. (SABMiller/Jason Alden/One Red Eye)**

**Above: Taking delivery of white maize at the Kitwe facility of Zambian Breweries. (SABMiller/One Red Eye)**

**Below: A truck returns to the depot after making deliveries of Nile Special. (SABMiller/Dave Parry/One Red Eye)**

of the multiple brewing industries, but some conclusions may be drawn. On a continent which was almost completely dominated by European colonial powers until the second half of the twentieth century, it is natural that the brewing traditions and the brewing companies of the colonial powers contributed to the development of present commercial brewing.

Given their size, it is little wonder that South Africa and Nigeria are the largest brewing countries on the continent. In the first decade of the twenty-first century, South Africa was producing around 26 million hectoliters annually, more than double its output in the 1980s before the end of Apartheid. Nigerian annual production, which stood at 11 million hectoliters in the 1980s, fell to 5.5 million at the end of the cen-

tury, but rebounded to 16 million a decade later. Other important brewing nations include Kenya and Cameroon, which average less than half of Nigeria's output.

Individual commercial brewing companies in Africa have long been at least partially owned by outside interests, with one exception. South African Breweries, Limited (SAB) was the largest brewing company in Africa for much of the twentieth century and the second-largest in the world since its acquisition of Miller Brewing in 2002. Now known as SABMiller, the company has also acquired smaller brands and brewing companies throughout the continent. Other major international companies with multiple breweries in Africa include Guinness (part of Diageo since 1997) and Heineken.

A beer traveler arriving in West Africa would notice Guinness, one of the post popular beers in the Ghana-Nigeria-Cameroon crescent, straight away. Indeed, the company has been successfully nurturing this market for decades. Guinness, which had been a global exporter of its products for more than a century, opened a brewery in Ikeja, near Nigeria's then-capital of Lagos, in 1962 and added another plant in neighboring Ghana in 1971. Two other Nigerian breweries were opened in Benin City in 1974 and Ogba in 1982. For many years, Guinness has sold more beer in Nigeria than in the United States.

The flagship Guinness product in Africa, as it is throughout much of the world, is 7.5 percent ABV Foreign Extra Stout. It is brewed locally using an "essence" produced in Ireland, combined with local fermentable grains. Unique among the Guinness Foreign Extra Stouts brewed elsewhere, the African breweries use maize and sorghum instead of barley. Launched in Nigeria in 2005, and also made with locally sourced grains, is 5.5 percent Guinness Extra Smooth, which is designed to approximate the flavor and character of the Draught Guinness familiar in Europe and North America. A similar non-alcoholic beverage called Malta Guinness, fortified with vitamins B and C, was introduced in Nigeria in 1990 and is now also made in Ghana, Cameroon, Cote d'Ivoire, Kenya, and Tanzania. Other Diageo products, ranging from Harp Lager to Satzenbräu Pilsner Lager, are brewed in Nigeria.

The largest brewing company in Nigeria, and all of West Africa, is Nigerian Breweries, a joint venture founded in 1946 by the British United Africa Company and Heineken. The first bottle of the company's flagship 5.3 percent Star Lager rolled off the bottling line in 1949. Additional breweries were added at Aba in 1957 and 2008, Kaduna in 1963, and Ibadan in 1982 as production grew steadily. The brewery expanded in the 1960s, building plants in

Chad, Ghana, and Sierra Leone. In addition to Star, the company has brewed various Heineken products as well as more of its own labels—Gulder Lager since 1970, Legend Extra Stout since 1992, and Goldberg lager and Life Continental Lager since 2011.

One of the largest single European possessions in Africa during the colonial period was the Belgian Congo (later Zaire and now the Democratic Republic of the Congo). With Belgium being the important brewing nation that it is, there is little wonder that brewing would take place here. In 1923, the Brasserie de Leopoldville was opened in the city of that name, now called Kinshasa, and through the years, additional brewing locations were opened in Boma, Bukavu, Kisangani, and Mbandaka. The company, owned by Groupe Lambert (later Bruxelles Lambert), became

**Above: Guinness Foreign Extra Stout had been brewed in Africa since 1962, and is one of the most popular beers in West Africa. (Diageo)**

**Below: Grain silos at the Zambian Breweries facility in Lusaka. (SABMiller/One Red Eye)**

Above: At work in the Cervejas de Moçambique brewhouse in Nampula, Mozambique. (SABMiller/Jason Alden/One Red Eye)

Below: Mozambique's popular 2M brand takes its name from the MacMahon Brewery, where it was first brewed. (SABMiller/Jason Alden/One Red Eye)

Bralima in 1957 and was nationalized by the Zaire government in 1975. Bralima struggled for a decade before the government allowed part of the seized shares to be returned to Bralima's former shareholders. It is now owned by Heineken and brews lagers such as Primus, Turbo King, and Mützig

as well as Guinness Foreign Extra Stout under license.

In the former French Congo, now the Republic of the Congo, the first commercial colonial brewery was established in 1952 in Stanleyville (now Brazzaville) under the auspices of Kronenbourg. A decade later, with independence, the Société Congolaise des Brasseries Kronenbourg (SCBK) was formed as the holding company. In 1968, the firm became Brasserie de Brazzaville under joint ownership of Heineken and Belgium's Bruxelles Lambert, with Heineken becoming the sole owner in 1984. Ten years later, the company merged with the Pointe-Noire Brewery to form Brasseries du Congo (BRASCO).

BRASCO brews foreign brands, including Heineken, but the long-time flagship lager for BRASCO and its predecessors is 5 percent ABV Ngok, the phonetic spelling of Le Choc, which means "crocodile." The Ngok was the symbol of the precolonial Ki-Kongo region of equatorial Africa, but the beer is a lager rather than a precolonial style.

An early European-style brewery in West Africa was the Overseas Brewery Limited, which began brewing in Accra in the British colony known as the Gold Coast (now Ghana) in 1931. The firm, later known as Accra Brewery Limited, became regionally famous for its Club Lager, which was popular throughout West Africa during World War II. In 1997, SAB acquired a nearly 70 percent share and a management contract.

Société Annonyme des Brasseries du Cameroon (SABC) was founded in 1948 by the French colonial brewing conglomerate Brasseries et Glacèries d'Indochine (BGI), famous for its 33 Export lager, widely available in Indochina and throughout Southeast Asia. BGI was bought in 1990 by the Castel Group, a French retailer with an interest in brewery acquisitions. BGI/Castel now has a 75 percent share in SABC, with Heineken owning a small share. In addition to 33, the product line is dominated by the brands of the parents instead of originals. Heineken's flagship lager as well as its Amstel brand are brewed by SABC, as are several French brands whose

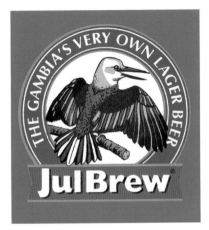

breweries in France have become part of the Heineken empire. These include Mützig, Pelforth, and Beaufort. In turn, these brands are also brewed at numerous other breweries across Africa, from Congo to Togo, in which Heineken owns a stake.

Meanwhile, unique indigenous national beer styles across West Africa include La Béninoise in Benin, Strela in Cape Verde,

Left: Banjul Breweries in Gambia use animals to identify each of their products. A kingfisher adorns the labels of their 4.7 percent ABV lager. (Banjul)

Above: In Abidjan, Solibra uses a crocodile to identify its Mamba lager, although a mamba is an African snake. Shown here is a Mamba export label. (Solibra)

Below: An overview of the packaging facility at Zambian Breweries in Lusaka, Zambia. (SABMiller/One Red Eye)

Above: Working on the St. Louis Lager bottling line at Kgalagadi. Breweries Limited at Gaborone in Botswana. (SABMiller/Philip Meech/ One Red Eye)

Below: A pair of St. Louis tap handles. The lager is the flagship brand of Kgalagadi Breweries. (SABMiller/ Philip Meech/One Red Eye)

Guiluxe in Guinea, Bière Niger in Niger, and both Eku Bavaria and Awooyo in Togo.

Banjul Breweries Limited in Gambia brew a range of products under the Julbrew brand name. Owned by a consortium headed by Deutsche-Amerikanische Handelsgesellschaft (DAH) and Bräuhaase International, the company started construction of its bottling plant in 1975 and began producing Gambia's only locally produced beer in 1977. These lagers, each identified on the label by an animal, include Julbrew Regular (kingfisher), Julbrew Export (crocodile), Julbrew Strong (hippopotamus), and Julbrew Very Strong (10 percent ABV), which has a cobra on the label.

Meanwhile, it is a crocodile, rather than another snake, which adorns the label of Mamba Lager,

brewed by the Solibra Brewery at its facility at Abidjan, the capital of Cote d'Ivoire.

In East Africa, the largest brewing company is East African Breweries Limited (EABL), which owns a 100 percent share of Kenya Breweries, 98.2 percent of Uganda Breweries, and 51 percent of Serengeti Breweries. EABL also owns other various distillers, vintners, and maltsters. In turn, EABL's own biggest shareholder is the Guinness parent, Diageo.

The company originated as Kenya Breweries, founded in 1922 by British expats George and Charles Hurst. The flagship brand is a lager called Tusker, which is so named to commemorate Hurst's having been killed by an elephant in the company's first year. Tusker is marketed under the slogan "Bia yangu, Nchi yangu," which is Swahili for "My beer, My country."

EABL also produces Guinness Foreign Extra Stout as well as a long series of lager brands including Allsopps, Bell, Pilsner, Pilsner Ice, President Extra (6.6 percent ABV), Senator, Serengeti, and White Cap. These brands as well as Tusker are available elsewhere in Africa, while Tusker is also sold outside Africa, notably in the United Kingdom and in the United States.

Another early European-style brewing company was Tanganyika Breweries Limited, founded in 1930 in Mwanza on the southern

shore of Lake Victoria, an area originally colonized by Germans. In 1964, with the independence and merger of Tanganyika and Zanzibar as Tanzania, the name changed to Tanzanian Breweries. The company was nationalized between 1974 and 1993, after which a 50 percent share was sold to SAB, who invested in brewery expansion, including new breweries in Moshi and Mbeya. The lagers brewed by Tanzanian Breweries include Kilimanjaro, Ndovu, and Safari.

Maputo, in the West African Portuguese colony of Mozambique, was home to the Laurentina Brewery, which began operations in 1932 with the launch of the Laurentina brand. Also in Maputo, the MacMahon Brewery was founded in 1965, quickly becoming famous for its popular 2M (Dois M in Portuguese) lager. Mozambique's most popular brand, 2M rates poorly in most international tasting comparisons. These breweries, as well as a third plant in Beira, were acquired by SAB in 1995 and merged as Cervejas de Mozambique. In 2011, the company introduced Impala, the first beer in Mozambique to be commercially produced using cassava.

In Uganda, Nile Breweries Limited were established in 1951 and began brewing in 1956 at Njeru, a suburb of Jinja, Uganda's second-largest city. The brewery is located adjacent to one of several

## THE MAN WHO WAS KILLED BY HIS TRADEMARK

In 1923, two brothers, British expatiates living in Kenya, went hunting. A year earlier, George and Charles Hurst had started a brewery called Kenya Breweries Limited, and were in the process of creating a lager. While on their hunting trip, poor George was killed by an angry bull elephant, which were then referred to in the vernacular as "tuskers."

To commemorate the event, Charles used this term to name the new lager, which went on to become one of the most popular brands in East Africa.

Today the brand has a 30 percent market share in Kenya, and is the sponsor of the Tusker Football Club. The classic, vintage label is shown above, while the contemporary label is depicted below. (Pictures: Diageo)

Above: A bottle of Mosi being pressure tested at the Ndola brewery in Zambia. (SABMiller/One Red Eye)

Below: Ndovu on the bottle line at Tanzanian Breweries Limited. (SABMiller/Jason Alden/One Red Eye)

competing locations held to be "the source of the Nile," and the brewery still draws its water from this source. Nile Beer, now 5.6 percent ABV Nile Special Lager, has always been the company's flagship brand. Other products are Club Pilsener, the highly regarded Nile Gold Crystal Malt Lager, and a sorghum beer called Eagle Lager. Nationalized between 1972 and 1992, Nile Breweries were acquired by Muljibhai Madhvani, who invited SAB to take a 40 percent share in 1997, and take over management. In 2001, SAB bought out Madhvani to become the majority shareholder.

Also in 2001, SAB bought a half interest in Chibuku Products Limited, the largest commercial producer of sorghum beer in Malawi.

In 2009, Southern Sudan Beverages Limited, a SABMiller subsidiary, began brewing White Bull Lager, the country's first locally produced beer, at new facility in Juba. SABMiller received funding from the Africa Enterprise Challenge Fund to "introduce a local sourcing model for cassava," which would be used in producing the beer, and which aided around 2,000 small farms in Sudan.

Another important brewing center in East Africa is Ethiopia, Africa's second most populous country. The major players are the government-owned Harar and Bedele breweries in the cities of the same name and the Meta Brewery in Abo. The former two were acquired by Heineken in 2011, who built a huge new plant in Addis Ababa, while Meta was purchased by Diageo in 2012. In addition to lagers bearing the brewery names,

Harar brews Hakim Stout, and Meta brews Guinness Foreign Extra Stout. Harar also produces non-alcoholic Harar Sofi for the Muslim market.

Brewed by the BGI-owned Kombolcha Brewery in Addis Ababa is a lager called St. George, named for Ethiopia's patron saint. Dating back to 1925, St. George is available at some Ethiopian restaurants in the United States and the United Kingdom.

Locally-brewed commercial brands, mainly lagers, that are associated with other countries across East Africa include Eritrea's Golden Star, Madagascar's Three Horses Pilsener, and Phoenix Lager, brewed since 1963 by Phoenix Beverages (formerly Mauritius Breweries), located on the island of Mauritius. Additional Phoenix products include Blue Marlin Lager, and the company has brewed and bottled Guinness Foreign Extra Stout under license since 1975.

While European-style brewing began on a significant scale in most of Africa in the twentieth century, it began in Southern Africa in the seventeenth century. In 1658, Jan van Riebeeck decided to start brewing in the Cape of Good Hope country near present-day Cape Town, but because yeast and hops had to be imported, his first beer was of marginal quality. The first professional brewer arrived at the Cape in 1694, and a

brewery was established on Simon van der Stel's farm in Newlands because it was described as having "the finest and best water available for this purpose." Today, the modern brewery at Newlands still uses water drawn from the same natural springs.

**Above: Enjoying Castle Lager, the long-time flagship brand of South African Breweries. (SABMiller/Jason Alden/One Red Eye)**

**Below: Inspecting Castle Lager on the bottling line at the Ndola brewery in Zambia. (SABMiller/ One Red Eye)**

**Above: Friends enjoying a couple of glasses of Laurentina on a warm Mozambique afternoon. (SABMiller/Jason Alden/One Red Eye)**

**Below: Safari is one of the lager brands produced by Tanzanian Breweries Limited. (SABMiller/Jason Alden/One Red Eye)**

Brewing remained a cottage industry until 1820, when Jacob Letterstedt established the Mariendahl Brewery at Newlands. In 1864, Anders Ohlsson arrived from Sweden, bought the Mariendahl Brewery, and built his Anneberg Brewery in 1883. Absorbing much of his competition, Ohlsson's Lion lager dominated the market and industry on the Cape. With the gold rush in the Transvaal came a beer rush, and Charles Chandler set up the first Transvaal brewery, Union Breweries, in 1887 in Ophirton. Chandler's major competitor was Frederick Mead, who built a large new brewery in Johannesburg. In 1895, Mead took on investors to form South African Breweries, Limited (SAB), and launched the iconic Castle lager brand. In 1899,

SAB entered the Cape market, purchasing the Martienssen Brewery in Cape Town and making inroads into Ohlsson's territory.

In the early twentieth century, SAB expanded to Port Elizabeth and Bloomfontein as well as to Salisbury, Rhodesia (now Harare in Zimbabwe). Because of its size, SAB and the Castle brand dominated the market across southern Africa, as it does to this day. SAB and Ohlsson coexisted profitably until the 1950s, when increased postwar taxation led to rising costs and a measurable decrease in consumption. In 1956, Ohlsson's holdings merged into SAB.

By the end of the twentieth century, on the eve of the merger with America's Miller Brewing, SAB was the leading brewing company on the African continent, with a near-total market share in the southern part of the continent.

The principal products in the SABMiller portfolio in South Africa include the signature 5 percent ABV Castle as well as 4.5 percent Hansa Pilsener and 5.5 percent Carling Black Label, which originated in Canada but is widely licensed throughout the world.

SABMiller also has ownership and management interests in smaller brewing companies across Africa that date back to the late twentieth century, when SAB was invited by various governments to play an equity and managerial role in the privatization of previously state-owned breweries. In Botswana this includes Kgalagadi Breweries and Botswana Breweries Limited, where SAB was invited to become the strategic partner in 1977. In 1994, the government of Zambia sold SAB an interest in Lusaka-based Zambian Breweries. Their leading brand is 4 percent ABV Mosi Lager, named for Victoria Falls, known locally as Mosi oa Tunya.

In 1995, two breweries and a Coca-Cola bottler in Matsapha, Swaziland, merged to form Swaziland Beverages Limited (SBL), which became a single-site subsidiary of SAB. The two breweries were Swaziland Breweries, founded in 1969 as a wholly-owned SAB subsidiary, and Ngwane Breweries, established in 1981 in Manzini. In 1980 and 1982, Lesotho Brewing and Maluti Mountain Brewery were founded as joint ventures between SAB and the Lesotho government, with the latter as majority shareholder. Between 2002 and 2010, they were privatized and merged as a SABMiller subsidiary.

Of the three multinational companies—Heineken, Diageo, and SABMiller—which control the lion's share of African brewing, only the latter is nominally African. Headquartered in Johannesburg when it was still SAB, it is now based in London.

## THE CHAIRMAN

**Ernest Arthur Graham Mackay had been with South African Breweries for more than two decades when he was named as SAB's CEO in 1999. On his watch, the largest brewing company in Africa became a world leader, then acquired America's Miller Brewing to become SABMiller, the penultimate brewer in the world.**
**Pictured below is the company's brewery at Port Elizabeth.**
(Photo above: SABMiller/Philip Meech; below: SABMiller)

# Asia

In the twenty-first century, China has emerged as the world's largest brewing nation—after having not even appeared in the global top 20 in 1980. Everything that has been written and said about the Chinese economy as a whole during this period is true about the Chinese brewing industry, except to add that much of the growth is for domestic consumption.

From 1930 through the 1949 revolution, there were fewer than ten commercial breweries operating in the entire country, with an annual production of between 100,000 and 200,000 hectoliters. With the upheaval of the Cultural Revolution and internal strife, little accurate data was available from the 1950s until the 1980s, but by 1985, China had approximately 300 breweries producing 16.5 million hectoliters. Within a decade, the number of breweries quadrupled. By the turn of the century, China was second in the world, brewing 200 million hectoliters. A decade later, the output had doubled to more than 400 million, twice that of the United States.

Since 2008, China also has the world's biggest beer brand. As reported by the Reuters news agency, 4.3 percent ABV Snow lager (literally Snowflake) eclipsed AB InBev's Bud Light with an annual "sales volume jump" of 19.1 percent. Snow is brewed by China Resources Snow Breweries (C.R. Snow), which is a joint venture between China Resources Enterprises and SABMiller dating

**Below: A still life with Snow lager at a sidewalk cafe in Beijing. Snow is the largest-selling beer brand in China, and in the world. (SABMiller/ Jason Alden/One Red Eye)**

to 1994. C.R. Snow operates more than 80 brewing plants, and has a better than 20 percent market share in China.

In 2011, C.R. Snow also acquired a controlling interest in Jiangsu Dafuhao Breweries and full control of the Shanghai Asia Pacific Brewery from Heineken. According to the company, C.R. Snow's other brands include such colorful names as Blue Sword, Green Leaves, Huadan, Largo, Löwen, New Three Star, Shengquan, Shenyang, Singo, Sip, Tianjin, Yatai, Yingshi, and Zero Clock.

The oldest modern commercial brewery still extant in China is the Harbin Brewery, founded in 1900 in the northeast city of the same name. Harbin was then evolving out of a small village at the terminus of the Russian-financed Chinese Eastern Railway. One the first Chinese breweries to be listed on the Hong Kong Stock Exchange, the brewery became the object of acquisition interest from both SABMiller and AB InBev

early in the twenty-first century. The latter was successful.

As the Harbin story illustrates, much of the contemporary brewing culture in China had its origins in the nineteenth century with the arrival of Europeans, who negotiated with the Qing Dynasty for leased enclaves in Chinese port cities. Among these, the United Kingdom leased Hong Kong, Portugal got Macau, and Germany took Tsingtao (now Qingdao).

Above: Serving up bottles of Snow lager, the world's leading beer brand, at a stylish, upscale establishment. SABMiller is a joint venture partner in C.R. Snow, the brewery behind this remarkable success story. (SABMiller/Jason Alden/One Red Eye)

Above: A classic label from Tsingtao, the Chinese beer brand best known in the West. (Author's collection)

Below: A selection of Chinese lager brands in which AB InBev is involved include Harbin of Harbin in Xiangfang, Jinling of Nanjing, and Sedrin of Fujian. (AB InBev)

Beer being important to the Germans, they invested heavily in a huge brewery in Tsingtao, which opened in 1903. It was taken over by the Japanese after World War I, nationalized after World War II, and privatized in the 1990s. Exported to the West since the 1970s, Tsingtao's flagship product, a 4.7 percent ABV lager, became well-recognized globally. It accounts for half of all Chinese beer exports. The company is still mainly owned by Chinese interests, with Japan's Asahi Breweries owning a minority stake. AB InBev also owned a minority share until 2009.

Along with Snow and Tsingtao, China's other two national brands are Yanjing and Zhujiang, both lagers. The Yanjing Brewery was started in the Chinese capital in 1980, was a major sponsor of the 2008 Olympics in Beijing, and is now one of the largest breweries in China. Yanjing is the brand name for several lagers of varying ABV as well as products such as the 4.3 ABV Yanjing Black dark lager. Established in 1985 in Guangzhou, in the Pearl River Delta of southern China, the Zhujiang (Pearl River) brand is owned by the Guangzhou Zhujiang Brewery Group, which owns two dozen related and unrelated enterprises as well as the Guangzhou Yongxin, Shijiazhuang, Zhanjiang, Dongguan, and Zhongshan breweries. The group's flagship beer is a 5.3 percent ABV lager called Zhujiang. The company is mostly state owned, although AB InBev has a minority interest.

AB InBev's other interests in China also include the Lion Brewery Group (formerly the Double Deer Brewing Group), based in Wenzhou, as well as the Jinling Brewery in Nanjing, acquired in 1997 by InBev's predecessor firm, Interbrew. Having acquired Harbin in 2004, the company bought the Sedrin Brewery in Fujian in 2006. In 2013, the *China Times* News Group reported that AB InBev had "struck a deal to acquire Nanchang Asia Brewery to help it tap into middle and low-end markets in China."

One of the more interesting twenty-first century beer marketing stories from China is a product called China Pabst Blue Ribbon, produced by CBR Brewing, which licenses the brand and logo from the American owners. Owned in the Virgin Islands, CBR operates from offices in Hong Kong and has brewing operations in Zhaoqing in Guangdong Province. Unlike the American variant, which is produced as a mass-market budget brand, CBR positions their Blue Ribbon as a high-status, super-premium brand, which the beer traveler will sometimes see on the shelves of specialty stores for around the same price by volume as moderately-priced spirits.

Though the immense Chinese brewing industry is heavily skewed to light lagers, many using rice as well as barley malt, the second decade of the twenty-first century

saw the door open slightly to craft brewing. Several American-style microbreweries and brewery restaurants opened in major cities. While they initially employed Western brewers brewing Western beer styles and catered mainly to the Western expat community, locals are discovering them, and Chinese brewers are coming on board. These include Sam Yang, who started at Henry's, the "first authentic microbrewery" in Shanghai, and who is now at the American-style Shanghai Brewery.

In addition to the Shanghai Brewery, a beer traveler visiting that city would have a choice of stopping by the Boxing Cat Brewery. Both of these operate at multiple locations. Boxing Cat's portfolio includes 4.5 percent ABV Right Hook Helles, 5.5 percent ABV Sucker Punch Pale Ale, 6.3 percent ABV TKO IPA, and 5.6 percent ABV Donkey Punch Porter, "a

**Above: In the lab at the Lion Brewery in Wenzhou. The brewery was previously known as the Double Deer Brewery. (AB InBev)**

**Below: China Pabst Blue Ribbon is a 3.3 percent ABV pale lager brewed under license by Guangdong Blue Ribbon in Zhaoqing. It is marketed as high-priced elite brand. (Guangdong)**

**Above: The South Korean pop star and television personality Psy (Park Jae-sang), famous for his 2012 hit single "Gangnam Style," was a spokesman for the Hite Brewery of Kangwon, South Korea. (Hite)**

**Below: Advertising for the Double Deer lager brand from Wenzhou, China. (AB InBev)**

**Right: The sensual pleasure of South Korea's OB lager. (OB)**

robust brown porter brewed with cacao and ancho peppers."

After Japan and China, South Korea is a distant third place among Asia brewing nations. Both based in Seoul, the largest brewing companies are Hite, founded as Chosun Breweries in 1933, and the Oriental Brewery Company (OB), which was started by the Doosan Group in 1952. OB was owned by InBev between 1998 and 2009. While Hite brews with a blend of barley and rice malt, OB uses rice exclusively. Hite's 4.5 percent ABV lager is the biggest-selling brand in South Korea. Other leading brands include 4.4 percent ABV OB Blue (formerly OB Lager) and Hite's 4.5 percent ABV Prime Max, which is brewed exclusively with barley malt. Hite also produces 4.5 percent ABV Hite Stout, which is based on Guinness Extra Stout. OB also brews 4.5

percent ABV Cass Lager, which was originally a product of the Jinro-Coors Brewery, which was absorbed by OB.

Because of laws which severely restrict retail activities by breweries, the microbrewery scene in South Korea is very limited. Exceptions are the American-style Craftworks Taphouse and Bistro in Seoul and the Craftworks Taphouse Pangyo in Bundang. Brewed "in small batches" by Kapa Brewery in Gapyeong, the products here include 4.5 percent ABV Namsan Pure Pilsner, 4.5 percent ABV Baekdusan Hefeweizen, and 6.8 percent ABV Jirisan Moon Bear IPA.

In the Philippines, the largest and best-known brand and brewery was also the first brewing company in Southeast Asia. From

operations which began in Manila in 1890, eight years before the Spanish relinquished control of the Philippines to the United States after the Spanish-American War, the San Miguel Brewery had always dominated the domestic market. Today, operating from a brewery complex at Mandaluyong City in the center of Metro Manila, it has a market share of around 90 percent. Its parent company, San Miguel Corporation, is Southeast Asia's largest publicly traded food and beverage company, with interests in other businesses from oil companies to airlines across Asia. In second place with around 10 percent, Asia Brewery was started in Makati City in 1982.

San Miguel's flagship lager, ubiquitous across the Philippines and well-known throughout Southeast Asia, is the 5 percent ABV San Miguel Pale Pilsen, a name which precisely describes the beer. Also brewed are San Miguel Premium All-Malt, San Miguel Premium Lager, and Cerveza Negra (San Miguel Dark Beer), all with an ABV of 5 percent.

In Southeast Asia, a number of important, large brewing companies founded early in the twentieth century have come under the same corporate umbrella in the early twenty-first century. Malayan Breweries, Limited (MBL) was formed in British-owned Singapore in 1930 as a joint venture between the Singapore-based

holding company Fraser & Neave and Heineken. The following year, Germany's Beck's started its subsidiary Archipelago Brewery in Singapore, brewing a lager called Anchor. Given its ties to Germany, Britain seized Archipelago in 1941 during World War II and sold it to MBL. In 1990, to reflect the company's regional market expansion, the name was changed to Asia Pacific Breweries Limited (ABPL or ABP).

In 1929, Heineken opened a brewery in Surabaya in the Dutch East Indies (now Indonesia). In 1949, it became Heineken's Indonesian Brewery Company. Between 1957 and 1967, the brewery was under Indonesian government control, but it reverted to Heineken as Multi Bintang Indonesia. In 2010, APBL acquired Bintang Indonesia from Heineken. Two years later, Heineken regained control once again, buying out all of the Fraser & Neave shares of APBL.

**Above: This illustration of the famous enamel label of 5 percent ABV San Miguel Pale Pilsen was commissioned for the brand's 100th anniversary in 1990. (Author's collection)**

**Below: Friends toasting with Jinling lager over dinner at a restaurant in Nanjing. (AB InBev)**

Above: Celebrating with Kingdom Pilsener at a night club in Phnom Penh, Cambodia. (Kingdom)

Below: A classic label for the legendary 5 percent ABV Tiger Beer, long the signature lager of Singapore. (Author's collection)

Below right: The bar at the Kingdom Brewery in Phnom Penh, Cambodia. (Kingdom)

The flagship product of MBL, and later of APBL, has been Tiger Beer, long marketed under the slogan, "It's Time for a Tiger." A popular 5 percent ABV lager, Tiger has been widely distributed throughout Southeast Asia since it was first introduced in 1932. In 1946, the writer Anthony Burgess used the slogan as a title for his book about life in Malaysia and Singapore. APBL also still brews Anchor Pilsener at 3.7 percent ABV as well as its 7 percent ABV ABC Extra

Stout. APBL maintains the independent brand identity of its Multi Bintang subsidiary, continuing to market Bir Bintang (Star Beer), a 4.7 percent ABV lager. As the parent company is now Heineken, the breweries also brew that company's own flagship lager.

In addition to the main breweries in Singapore and Indonesia, Heineken-owned APBL also operates breweries in Cambodia, China, Laos, Malaysia, Mongolia, New Caledonia, New Zealand, Papua New Guinea, the Solomon Islands, Sri Lanka, Thailand, and Vietnam.

From MBL, APBL inherited an interest in the Guinness Anchor Berhad (GAB) Brewery at Selangor on the west coat of Malaysia, north of the capital of Kuala Lumpur. Founded by Guinness in 1964 as the company's first Asian brewing venture, the brewery was originally called Guinness Malaysia Limited and later Guinness Malaysia Berhad. GAB was formed in 1989 with MBL acquiring a share in the brewery. Both APBL products, such as Tiger, and Guinness

Foreign Extra Stout are among the products produced here.

Unlike many cities in the Far East, the beer traveler will find that Singapore also has a robust brewpub and microbrewery scene. One of the first was Brewerkz, with several locations, including Riverside Point and the Singapore Indoor Stadium. Accounting for three quarters of sales in the highly rated 4.5 percent ABV Brewerkz Golden Ale. Other products include a 6 percent ABV IPA, 8 percent ABV Blind Pig Stout, and Belgian-style Moh Gwai at 7.2 percent ABV. Since 2013, Brewerkz has exported its beer to Bangkok, Thailand.

Other brewery restaurants in Singapore include the Red Dot Brewhouse, the Pump Room, and a satellite operation of the New Zealand–based Moa Brewery. Using the Archipelago brand name, APBL operates a number of "concept" brewpubs across Singapore, which brew small quantities of specialty beer. In the penthouse of the Marina Bay Financial Centre, the beer traveler will find LeVeL33, called "the World's Highest Urban Craft-brewery."

In Thailand, Singha has been the leading brand since it was first brewed here in 1933. It originated with Boon Rawd Sreshthaputra, an ambitious young man who took an interest in giving Thailand its first modern commercial brewery and traveled to Europe to see how it was done. He returned to set up

his Boon Rawd Brewery in 1933 and named his beer Singha after the mythical lion of Hindu mythology. The original Singha is a 5 percent ABV lager, and there is now a 3.3 percent ABV Singha Light.

Singha's biggest competitor in Thailand is Chang Beer, which was first introduced in 1995 by Thai Beverage (ThaiBev). Founded by Charoen Sirivadhanabhakdi, the ThaiBev company also has global holdings in the spirits business, and it has owned Fraser & Neave since 2013.

**Above: The BeerVault at Four Points by Sheraton, at Sukhumvit 15 in Bangkok, serves nearly 100 international and local beers. (Starwood Hotels & Resorts)**

**Below: An export label for Thailand's famous 5 percent ABV Singha lager. (Author's collection)**

**Above: Cambodia's Kingdom Pilsener poses with a Buddhist temple. (Kingdom)**

**Below: A classic label from Taj Mahal Lager. It is brewed at various locations throughout India by Bangalore-based United Breweries (UB), and is a widely exported brand. (Author's collection)**

Phuket Lager, which first appeared in 2002, is another Thai brand made with rice malt. Named for the southern Thailand resort city that is so popular with Western tourists, Phuket is specifically marketed to "young 18-30-year-olds familiar with the beauty and tranquility of Phuket as an escape destination or paradise island called home. Perhaps the socially adventurous, in search of a sophisticated party atmosphere."

Phuket is financed through Leopard Capital, a private equity firm with holdings throughout Southeast Asia that included Cambodia's Kingdom Breweries, which were, in the company's words, "established in 2009 with the vision of becoming Cambodia's first internationally recognized premium beer brewer."

Echoing the demographic that Leopard sees for Phuket in Thailand, the company goes on to explain that "in recent years, Cambodia has emerged as one of Southeast Asia's premier travel destinations, hosting world-class heritage sights, a burgeoning nightlife, and pristine beaches," adding that "Phnom Penh has transformed from a sleepy city into a bustling urban area in recent years; tourists, expats and young Khmers alike explore the plethora of bars and bistros lining the city's riverside district." Kingdom produces 5 percent ABV Kouprey Dark Lager and 5 percent ABV Kingdom Gold Lager.

Cambodia's best-selling brand, 5.5 percent ABV Angkor Beer, is named for the Angkor Wat temple complex and brewed by the Cambrew Brewery in Sihanoukville. The brewery also produces 6 percent ABV Klang Lager, and two 8 percent ABV stouts, Angkor Extra and Black Panther Premium.

Beer, notably India Pale Ale, came to India with the British in the eighteenth century, but brewing was slow to follow. There were a few small- to moderate-sized breweries established in India but no large scale effort on the part of the British brewers to establish a brewing industry within India. Edward Dyer established the first English brewery in India in the 1820s at Kasauli in the Himalaya foothills, where he could take advantage of abundant clean, fresh water to brew his beer, called Lion, and cater to the English expats who summered in the mountains. Eventually, he built several breweries across India.

Dyer later moved into whiskey distilling and sold his breweries at Shimla and Solan to H.G. Meakin, who built additional breweries under the Dyer Meakin Breweries Limited name. In the late 1940s, this company was acquired by N.N. Mohan, who built additional breweries and renamed the firm Mohan Meakin Breweries in 1967. With headquarters in Chennai (formerly Madras), the company moved into spirits as well as beer, and the name evolved to Mohan Breweries & Distilleries Limited. Today, Mohan still brews a brand called Lion, although it is now a lager, not an ale. The company's flagship Golden Eagle 4.8 percent ABV lager is one of India's biggest selling brands.

In 1857, the Castle Brewery was established in southern India to supply the plantations spread across the Nilgiri Hill Country. The Bangalore Brewery Company was set up in 1885 to serve British troops based in the Bangalore area, and the British Brewing Corporation was established in the port city of Madras in 1902. In 1915, a Scotsman named Thomas Leishman consolidated these firms into the company that grew into today's largest Indian brewing company, United Breweries Limited (UBL).

In 1947, when India gained its independence, the UBL headquarters moved to Bangalore, and Vittal Mallya, who would play a key role in UBL's development, joined the board, later to serve as chairman. In 1983, he was succeeded as chairman by his son Vijay Mallya. It was he who transformed UBL, now the UB Group, into India's largest brewer and into a global and diversified conglomerate, active in the beer and spirits market and well as other industries from real estate to Kingfisher Airlines. Heineken currently owns about a third of UB's shares. The company's flagship is the Kingfisher family of lager brands, market leaders in India, ranging from 4.8 to 5 percent ABV. There is also 7.1 percent Kingfisher Strong. Other lager brands in the UB product portfolio are 4.5 percent Taj Mahal and 4.7 percent Flying Horse.

**Above: A waiter serves a tray of 5 percent ABV Royal Challenge lager in a restaurant in Mumbai, India. (SABMiller/Philip Meech/One Red Eye)**

**Below: Enjoying cold mug of 7 percent ABV Haywards 5000, one of India's leading strong lager brands. Like Royal Challenge, the brand originated with from the Shaw Wallace Company, was briefly brewed by United Breweries (UB), but is now part of the SABMiller India portfolio. (SABMiller/Philip Meech/One Red Eye)**

# Japan

Among the world's top ten brewing nations since the latter half of the twentieth century, Japan is home to some of the world's big lager conglomerates while also having Asia's most robust craft brewing scene. Microbreweries, called ji-biru, which roughly translates as regional brewery, have proliferated since 1994, when it became legal for a brewery to brew *less* than 20,000 hectoliters annually.

Japan's Craft Beer Association sponsors a number of events across Japan through the year. These include the Great Japan Beer Festivals in Tokyo, Osaka, Nagoya, and Yokohama as well as the Asia Beer Cup (formerly the Japan Asia Beer Cup) and the Nippon Craft Beer Week, held in Tokyo in June.

As with major brewing companies in most places in the world, the beer brewed by Japan's major brewers is predominantly lager. Within that category, there are a variety of types classified by whether they contain all, some, or no malted barley. As with American mass-market lagers, most Japanese lagers contain a proportion of malted rice. A category called rice lager contains a very large proportion of rice, and happoshu, a distinctive Japanese style, contains mainly rice. As it is not called beer, it is taxed at a lower rate than beer. Happoshu is quite popular in Japan, though in recent years market share is

Below: Inside the brewhouse at the Kiuchi Brewery in Naka, Ibaraki Prefecture. Still a microbrewery by comparison the Japan's megabreweries, Kiuchi has been brewing since 1823. (Sadamu Saito, Kiuchi)

dropping off. Most major Japanese brewing companies are also heavily involved in the international spirits business.

Western beer was first introduced into Japan in the seventeenth century by the Dutch, who had done the same in what is now the United States. Sake, which is brewed by a process similar to that of beer, was produced in Japan by the eighth century, but modern Western brewing was not undertaken on a significant scale until

the second half of the nineteenth century.

After the Meiji Reformation of 1868, the first successful commercial brewery was set up in 1869 at Yokohama by the American firm of Wiegand & Copeland. According to sources, including Lynn Pearson, in the book *Towers of Strength*, this company was later sold to Japanese interests and evolved into the Kirin Brewery Company. According to Kirin, its forerunner was Japan Brewery

This page: Kirin is Japan's largest brewing company, with extensive holdings outside the country from China to Australia to Brazil. It owns an interest in San Miguel in the Philippines, and in the operations within Japan of Heineken and Coca-Cola. Kirin is also active in wine, spirits and soft drinks. The "kirin" on the label is a creature common in Asian mythology as a symbol for good fortune. Part dragon and part deer, it is often reported to have been sighted on the occasion of the births of illustrious persons. Legend holds that in the sixth century BC, a kirin was seen by the mother of Confucius shortly before his birth. (Kirin)

Above: Classic labels from Sapporo. On the left, Sapporo Draft Beer is a 5 percent ABV rice lager, while the products labeled as lagers are traditional lagers. (Author's collection)

Below: As can be seen in the glass, 5 percent ABV Asahi Black (Kuronama) is not actually a black lager, but the equivalent of a Munich dunkel. The major Japanese breweries have long included such "black" beers in their portfolios. (Asahi)

Company Limited, established on the site of the former Spring Valley Brewery in Yokohama in 1885. The company was founded by "W.H. Talbot, E. Abbott, and a few other foreigners residing in Yokoyama, all of whom held a conviction that beer would become popular with Japanese people. On the advice of Thomas Blake Glover, who was a Scottish merchant and who would later join the board of Japan Brewery, Yanosuke Iwasaki—president of Mitsubishi—and eight other Japanese bought interests in the company." The company became the Kirin Brewery Company Limited in 1907.

Today, Kirin is the largest brewing company in Japan. The third largest in the world in 1980, and the largest in Asia until the turn of the century, Kirin still ranks in the global top ten. Owned by Mitsubishi, Kirin is a diversified multinational with interests in spirits and soft drinks as well as beer and with holdings in pharmaceuticals, real estate,

and other products and services. On the brewing side, Kirin has brewing operations throughout Japan and owns several overseas breweries, including Lion Nathan in Australia, Taiwan Kirin in Taiwan, Zhuhai Kirin in China, and Brasil Kirin (formerly Schincariol). Kirin also has an interest in Singapore-based Asia Pacific Breweries and San Miguel in the Philippines.

Lagers brewed by the company include 6 percent ABV Kirin Deluxe and 5 percent ABV Kirin Ichiban. Classified as Japanese-style rice lagers are 5 percent Kirin Ichiban Maribana, 4.5 percent Kirin Lager, and 4.5 percent Kirin Classic Lager. Kirin Ichiban Shibori Stout is not a stout, but a 5 percent schwarzbier (black lager). Kirin has several happoshu brands which range between 4 and 6 percent ABV.

Sapporo Breweries were the second of Japan's Big Four brewing companies to be founded. Though now headquartered in Tokyo, the company began operations as the Kaitakushi Brewery in the city of Sapporo on the northern island of Hokkaido in 1876. In 1906, it merged with the Japan Beer Brewery Company, which was founded in 1887 in Tokyo and brewed Yebisu Beer. Because of competition with Kirin, these two companies also merged with Osaka Brewing (the forerunner to Asahi) to form the Dai-Nippon Beer Company. After World War

II, Dai-Nippon divested Asahi, and in 1956 the company was renamed Sapporo after its birthplace and one of its popular brand names. Today, Sapporo operates the Hokkaido Brewery in Eniwa, the Chiba Brewery in Chiba, the Kyushu Hita Brewery in Hita, the Sendai Brewery in Natori, and the Shizuoka Brewery in Yaizu, all of which were built since 1971.

Most of Sapporo's beer portfolio is rated at or around 5 percent ABV, an exception being Yebisu Cho-choki-jukusei, a 6 percent lager. Among the 5 percent lagers are Sapporo Classic and Yebisu Premium. Black lagers, also at 5 percent, are Yebisu Black Beer and Yebisu Creamy Top Stout. Two 5 percent rice lagers include Sapporo Draft and Sapporo Premium. The happoshu line includes Sapporo Draft One and Sapporo Mugi To Hoppu (Barley and Hops), both at 5 percent.

Asahi Breweries, founded in Osaka in 1889 as the Osaka Beer Company and later part of Dai-Nippon, became independent after 1956 and grew into a major player in the Japanese market. Asahi has interests in brewing companies around the world, including shares in Australia's Foster's Group and China's Tsingtao. Until 1987, the flagship brand was Asahi Gold, but since then it has been the 5 percent ABV Asahi Super Dry, described as "Japan's first dry beer [which] set a new standard that is acknowledged

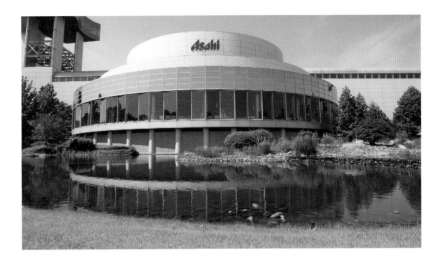

around the world today." In fact, it helped to inspire the "dry beer" craze that was popular among American mass-market breweries in the 1990s.

Other leading members of the portfolio are Asahi Jukusen, a 5 percent ABV rice lager, and Asahi

**Above: Headquartered in Tokyo, Asahi has a 40 percent market share in Japan. (Asahi)**

**Below: Mitsuboshi offerings from the Land Beer Brewery in Aichi-ken Nagoya-shi include 5 percent ABV Pale Ale, 5.5 percent Pilsner, and 6 percent Vienna-style lager. (Land Beer)**

Above: Stirring the brew at the Kiuchi Brewery in Naka. (Sadamu Saito, Kiuchi)

Below: From Kiuchi, 5 percent Hitachino Nest White Ale, and 7 percent Hitachino Nest Red Rice Ale. The latter is not a rice beer, but a Belgian-style ale. (Kiuchi)

Kuronama, a 5 percent dark lager. The 8 percent Asahi Stout, unlike other Japanese beers called stouts, actually is a stout. Asahi also produces several 5 percent happoshus and Asahi Strong Off, a happoshu rated at 7 percent.

In 1923, Shinjiro Torii built the first whiskey distillery in Japan at Yamazaki near Kyoto, thus founding the Japanese whiskey industry. His company, Suntory, was founded in 1899 with the intention of developing wine for export, and beer was not added to the Suntory product line until 1963. Suntory is really more of a vastly diversified real estate/leisure services company than a beverage producer and more of a spirits company than a brewer. Nevertheless, Suntory, along with Asahi, is among Japan's top four brewing companies. Suntory lagers include Malt's at 5 percent ABV, Premium Malt's at 5.5 percent and Premium Malt's Kuro, a 5.5 percent black lager. Malt's Bitter is an ale.

In distant fifth place among Japanese brewing companies, Okinawa-based Orion Breweries Limited has only 1 percent of the total Japanese beer market, but half the market on the island. The company began brewing in 1959, catering to the many American troops stationed on Okinawa. The flagship beer is a 4.7 percent ABV rice lager called Orion Premium Draft Beer.

The Kiuchi Brewery at Naka in Ibaraki Prefecture is an example of a Japanese microbrewery that has been around since 1823, producing both beer and sake. Founded by Kiuchi Gihei, the brewery was essentially a farmhouse brewery until the 1950s when the rapid expansion of the Japanese economy increased demand for beer and sake. In 1996, Kiuchi began marketing a range of craft beers with a distinctive owl character logo under the name Hitachino Nest Beer. In 2003, they began production of Shochu Kiuchi, a distilled liquor made from a by-product of sake brewing.

The Kiuchi Nest Beer variants include 5.5 percent ABV Pale Ale, 4.5 percent Sweet Stout, 5.5 percent Belgian-style White Ale, 7.8 percent Red Rice Ale, 8 percent Commemorative Ale, and 6.5

percent Nipponia, which is brewed using the revived Japanese breed of Kanego Golden barley, which was first developed in 1900, along with another strain of hops called Japanese-bred Sorachi Ace. Most of the product line is exported to the United States.

One of the most highly regarded of Japanese micro-breweries is the Baird Brewing Company founded in 2000 by Bryan Baird and his wife Sayuri in Numazu, Japan. As they describe themselves, "We are a family company born of a deep passion for beer and a great reverence for brewing history, tradition and culture. Our motto is 'Celebrating Beer.'" Baird Brewing products are marketed through an expanding circle of pubs, restaurants, and liquor stores as well as through proprietary pubs such as the Numazu Fishmarket Taproom in Numazu and the Nakameguro and Harajuku Taprooms in Tokyo.

Notable Japanese brewery restaurants include Kinshachi, the London Pub Towser, and Moku Moku, all in Nagoya, as well as Anjo DenBeer in Anjo, Loreley in Inuyama, Satsuma Brewery in Kagoshima, Shirayuki in Hyogo Prefecture, and the Yamato Brewery in Nara. For the beer traveler visiting the city of Sapporo, there is the Sapporo Beer Garden, with two large restaurants, a beer hall, and beer brewed on premises. It is adjacent to the Sapporo Beer Museum. In Tokyo, the beer travelers seek out the Koenji Bakushu Kobo and Swan Lake brewpubs.

**Above: The Kiuchi Brewery in Naka by night.** (Sadamu Saito, Kiuchi)

**Below: Kiuchi's Nipponia, a 6.5 percent ABV double IPA.** (Kiuchi)

# Australia and New Zealand

Below: Enjoying a warm day at the beach in the company of 3.5 percent ABV Pure Blonde Lager and 4.9 percent Crown Lager, both brewed by Carlton & United. (SABMiller/Tom Parker/One Red Eye)

Through consolidations which have taken place since the latter decades of the twentieth century, the number of major breweries in Australia has declined, and these are sorted into three holding companies as detailed below. Meanwhile, there has been a proliferation of smaller craft breweries which have come on the scene since the mid-1990s, most of them since the turn of the twentieth century. Many have come and gone, but about two dozen existed at press time, with more in the offing.

The first of the Big Three holding companies is Lion Nathan, which owns Castlemaine Perkins of Queensland, dating to 1878; J. Boag & Sons (formerly Esk) of Tasmania, dating to 1881; South Australian Brewing of South Australia, dating to 1859; Swan Brewery of Western Australia, dating to 1837; and Tooheys of New South Wales, dating to 1869. The "Lion" in the name originated in New Zealand with New Zealand Breweries, which was founded in 1923, became Lion Breweries in 1977, and became part of the expanding holding company, Lion Nathan, in 1988.

The second of the larger holding companies is the Foster's Group (formerly Elders IXL), which grew out of Carlton & United Breweries (CUB) of Victoria and Queensland, dating back to 1907, and which acquired the Cascade Brewery in Tasmania, dating to 1824.

Finally, the third holding company consists only of Coopers Brewery of Adelaide in South Australia, dating to 1862. Coopers has but a 5 percent share of the total market. The Big Three is really Coopers plus a Big Two that controls virtually all of the Australian market.

Of the Big Three, Lion Nathan is owned by Japan's Kirin and Foster's by SABMiller, while Coopers is still controlled by the Cooper family despite several takeover attempts by Lion Nathan. With the exception of Foster's, most Australian brewing company names and brands which are possessive words ending in "s" are written without an apostrophe.

The leading brands in Australia are XXXX from Lion Nathan and VB (Victoria Bitter) from Foster's, each with about a 12 percent market share. In a second tier, and just below 10 percent, are Foster's Carlton Draught and Tooheys New from Lion Nathan.

Although beer had arrived in Australia on ships from Britain as early as 1788, it deteriorated during the long voyage. In 1796, a settler named John Boston was the first to make commercial beer in Australia, using malted maize and Cape gooseberries.

Today, the most common type of beer in Australia is lager, but unlike the rest of the world, lager

**Above:** Tap handles at this Brisbane pub include Carlton Draught and Dry, as well as Carlton's Pure Blonde and 4.7 percent ABV Fat Yak Pale Ale from Matilda Bay Brewing, Australia's first craft brewery. (SABMiller/Tom Parker/One Red Eye)

Above: A bottle of Carlton's Crown Lager is flanked by Premium Lager and Pure All-Malt Lager from Tasmania's Cascade Brewery. (SABMiller/Tom Parker/One Red Eye)

Below: The Cascade Brewery on Cascade Road outside Hobart, Tasmania is arguably Australia's most picturesque brewing site. (Author's collection)

did not make an impact on production until the 1880s. The hot weather encouraged beer drinking, but it also played havoc with yeast. Ales were made, although, in most cases, not well. "Tasteless, insipid, and sugary" was the verdict on locally made beer in the early days of colonial settlement.

Coopers, Australia's oldest brewing company, was started by Thomas Cooper in 1862 by accident when he created his first batch of beer as a tonic for his ailing wife Ann. During the twentieth century, as domestic and export sales grew, the family continued reinvesting in plant and equipment. This culminated in the new Regency Park Coopers Brewery, opened in 2000. Today, the family-owned company still has a half dozen Coopers in its management, including Dr. Tim Cooper, the managing director. Coopers produces a range of lagers ranging in ABV from non-alcoholic (0.5 percent) to 5 percent. Ales range from 3.5 to 7.5 percent and include Coopers Original Pale Ale and Dr. Tim's Traditional Ale, both at 4.5 percent ABV. Also top-fermented is 6.3 percent Coopers Best Extra Stout. In addi-

tion to distributing its own beers, Coopers also markets an extensive range of homebrewing kits.

Castlemaine Perkins originated with Nicholas and Edward Fitzgerald, the sons of Francis Fitzgerald, a well-known Irish brewer. The brothers first came to Australia in the 1850s and established the Castlemaine Brewery in Milton, near Brisbane, in 1878. Their "Castlemaine (XXXX) Sparkling Prize Ale" won a Special Prize at the First International Exhibition in Sydney in 1883, but the now iconic brand four Xs was not widely used until the twentieth century.

In 1924, the familiar XXXX Bitter Ale was introduced. The advertising included a cartoon of a smiling, winking man wearing a boater hat with four Xs. Quickly dubbed "Mr.. Fourex," he has been a symbol of the brewery ever since, just as beers with the XXXX brand and remained among Australia's favorite beers.

The company became Castlemaine Perkins through the acquisition of the Perkins Brewery in 1928. In 1980, Castlemaine Perkins merged with Tooheys to form Castlemaine Tooheys. Tooheys company originated with Matthew and Honora Toohey, who emigrated from Ireland in 1841. They settled in rural Victoria to raise cattle, but they moved to Melbourne in 1860 to run the Limerick Arms Hotel. In 1865, on

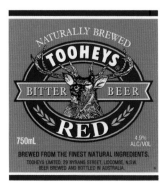

a visit to Sydney, their son John Thomas Toohey heard that the Darling Brewery in Sydney was for sale and convinced his brother James Matthew to join him in the business. They began brewing ale in 1869 and added a lager, Tooheys New, to their portfolio in 1902. Though not used on all products, the "stag" logo has been part of the history of Tooheys since 1869. The stag can be traced to John Toohey's favorite hotel, the Bald Faced Stag Hotel in Leichhardt. The head of a stag mounted on the bar wall caught Toohey's attention, and folklore holds that there and then he decided to use the stag as the symbol for the new brewery.

Castlemaine Tooheys was acquired in 1985 by Australian billionaire investor Alan Bond, who had also purchased breweries in the United States. In the 1990s, with Bond having gone bankrupt and facing a jail sentence, his empire was unwound. Castlemaine Tooheys was gradually acquired by Lion Nathan between 1990 and 2007.

Another part of the Lion Nathan Australian portfolio, J. Boag & Son was founded in 1881 on the Esk River in Tasmania as the Esk Brewery by James Boag I and was renamed in 1883. It remained family owned and operated until great-grandson George Boag retired in 1976. In 2000, it was acquired by the San Miguel Corporation of the Philippines, and sold to Lion Nathan in 2007.

Today, the Castlemaine-branded Lion Nathan products include the XXXX line of pale lagers, which range from the 3.5 percent ABV XXXX Gold to 4.8 percent XXXX Bitter. The Toohey's pale lager brands include the popular 4.6 percent ABV Tooheys New and 5.2 percent Tooheys Extra Dry Platinum. Also in the line is 4.5 percent Tooheys Red, a dark lager, and 4.8 percent Tooheys Old, a dark ale. A line of pale lagers are brewed in Tasmania under the James Boag name, ranging from 2.9 to 4.8 percent ABV. Also brewed in Tasmania are 4 percent Tasman Bitter and 5

**Above:** Classic labels from the classic breweries that became parts of the Lion Nathan portfolio include those of Brisbane's Castlemaine Perkins; Tooheys of Lidcombe, New South Wales; Swan of Canning Vale, Western Australia; and J. Boag & Sons (the Esk Brewery) of Launceton, Tasmania. (These images from the author's collection were supplied by request by the breweries in the 1990s)

**Below:** Created in 1924 to promote Castlemaine's XXXX Bitter Ale, the smiling, winking "Mr.. Fourex" still remains the most recognized spokesman for a beer brand in Australia, just as XXXX remains one of Australia's favorite beer brands. (Author's collection)

Above: A range of lagers that are part of the SABMiller (formerly Foster's) portfolio include 4.5 percent ABV Carlton Dry; 4.9 percent Crown Lager; 2.8 percent Cascade premium Light; and 4.9 percent Bluetongue Premium Lager.

The Bluetongue Brewery was a craft brewing company founded in 2001 in Hunter Valley, New South Wales, and subsequently sold to SABMiller, who later opened a new brewery in Warnervale to produce Bluetongue brands. Australia is home to the blue-tongued lizard (*Tiliqua adelaidensis*), as well as the Aboriginal legend of the mythical Bluetongue Lizard, a trickster and sorcerer. (SABMiller/Tom Parker/One Red Eye)

percent Wizard Smith's Ale. Lion Nathan also holds the licenses to brew Beck's, Guinness, Heineken and other international brands in Australia for sale in Australia.

The Swan Brewery originated with Frederick Sherwood, who began brewing in Perth in 1857. Captain John Maxwell Ferguson and William Mumme of the Stanley Brewing Company leased the Swan Brewery in 1874 and around 1887 acquired the Lion Brewery. In 1908, the name of the Stanley Brewery was changed to Emu Brewery, and in 1928 Swan purchased Emu. Owned by Alan Bond during the 1980s, Swan became a wholly owned subsid-

iary of Lion Nathan in 1992. The Swan product line then included those represented by the sign of the graceful swan as well as those of the flightless emu, a bird native to Australia. Lion Nathan eventually retired the Swan brand, but continued to brew a range of Emu lagers with ABV between 3.5 and 4.6 percent.

Carlton & United Breweries (CUB), which evolved into the Foster's Brewing (formerly Elders IXL) Group, had its roots in the old Carlton Brewery, born during the 1850s when the city of Melbourne had 35 breweries. CUB itself was born in 1907 through a mass merger that included the Carlton, Foster's, McCracken, Shamrock, and Victoria breweries. Over the next quarter century, CUB acquired the Abbotsford Co-operative, Ballarat, Cairns, McLauchlin, Northern Australian Breweries, Queensland Brewery, Richmond Brewery, and Tooth & Company. Between 1999 and 2011, the company also owned Samoa Breweries in Samoa.

In 1983, CUB was bought by Elders IXL, renamed in 1990 as the Foster's Group, after the most internationally famous product, Foster's Lager. This is how Foster's went from being a subsidiary of CUB to its parent. Foster's is named for a pair of American brothers who spent only about 18 months in Australia. They started their brewery in Melbourne in

1887, sold it the following year, and went home, never to be heard from again. Today, Foster's is better known abroad than at home.

The international flagship brand is represented in the product line by Foster's Lager, which is marketed around the world with a ABV that ranges generally between 4 and 5.2 percent, depending on the country in which it is found. The company also produces a number of lagers in the 3.5 to 4.9 percent ABV range under the Carlton name. These include Carlton Cold, Carlton Dry, and a family of citrus-infused beers. Also produced is a 4.9 percent ABV stout called Carlton Black and 5.7 percent Sheaf Stout, a long-time favorite in New South Wales.

The best-selling Foster's brand in Australia, VB (Victoria Bitter) is a 4.6 percent ABV lager. The family of brands includes 3.5 percent ABV Victoria Bitter Midstrength and 4.5 percent VB Raw Dry. Also produced is 4.6 percent Melbourne Bitter.

Meanwhile, the venerable Cascade Brewery in Tasmania produces lagers such as Cascade Lager and Cascade Blonde, both 4.8 percent ABV, and two 5.8 percent stouts, Cascade Export Stout and Cascade Special Stout, as well as 5.2 percent Cascade Pale Ale.

As we have seen, the nineteenth century was a period of entrepreneurship in Australian

## Beer Festivals in Australia and New Zealand

Australia boasts a full calendar of beer festivals, including the **Sydney Oktoberfest**, held in suburban Cabramatta. Also in Sydney in October is the annual **Australian Beer Festival**, hosted by the Australian Heritage Hotel, during which Gloucester Street is closed off for a weekend of fun. Brisbane and Canberra both observe **Oktoberfest,** and the **Canberra Craft Beer Festival** takes place in April. In November, it is time for the **Bitter and Twisted Boutique Beer Festival** at the Maitland Goal, the notorious former prison in East Maitland, New South Wales. In January, beer festivals occur in Tamar Valley, Tasmania and Ballarat, Victoria. In February, Perth celebrates **FeBREWary** and the **Taste Of The World Beer Festival** comes to Newcastle in January. Annual **Beer Week** festivities take place in Melbourne in May, in Western Australia in June, and in Queensland in July.

In New Zealand, the **Heartland Beer Festival** takes place in Queenstown in February, while March is a big month on the beer festival calendar. Events include the **New Zealand Beer Festival** in Auckland, the **Great Kiwi Beer Festival** in Christchurch, and the **Greater Wellington Brewday.** Later in the year, Wellington and Urenui both celebrate **Oktoberfest**.

brewing, while the first seven decades of the twentieth century were marked by closures and consolidations. In 1983, against this backdrop, and with echoes of the craft brewing revolution in America wafting across the Pacific, Phil Sexton, Garry Gosatti, John Tollis, and Ron Groves formed a company called Brewtech, which evolved into the first new brewery in Australia since before World War II. Brewing small batches at a series of locations around Perth in Western Australia, the company became the Matilda Bay Brewing Company, which was acquired by

Below: Foster's is the most popular Australian brand in the export market. (SABMiller/Tom Parker/One Red Eye)

**Above: Dating to 1862, Cooper's of Regency Park, near Adelaide in South Australia is the oldest and largest independent brewing company in Australia. (Author's collection)**

**Below: Dating to 1984, Matilda Bay Brewing was Australia's first craft brewery. It was acquired by Carlton & United, which was in turn acquired by SABMiller. (SABMiller/ Tom Parker/One Red Eye)**

the Foster's Group in 1990. A second brewery was established in metropolitan Melbourne in 2005, and brewing in Perth was terminated in 2007. Foster's then moved some Matilda Bay production to their Cascade Brewery in Hobart, Tasmania.

The larger brewing companies in Australia have moved more aggressively than their American counterparts to acquire craft breweries. In 2007, Foster's, now part of SABMiller, acquired the Bluetongue Brewery in Warnervale, New South Wales, brewers of the Bondi Blonde brands. Lion Nathan, meanwhile, owns Sydney's Malt Shovel Brewery, started in 1998, as well as the Little Creatures Brewery, founded in 2000 in Fremantle, Western Australia.

The Gage Roads Brewing Company in Palmyra, near Perth, is named for an anchorage near the port of Fremantle. The com-

pany was founded in 2002 by John and Bill Hoedemaker, with Bill having been one of the brewers at Matilda Bay. The company is one of Australia's largest and fastest-growing craft brewers, and has medaled at both the World Beer Cup and the Australian International Beer Awards. Gage Roads beers include 4.7 percent ABV Gage Roads Premium Lager, 7.4 percent Abstinence Belgian Dubbel Chocolate Ale, and the award winning 5.4 percent Sleeping Giant IPA.

Among the other craft brewers is John Stallwood's Nail Brewing, which began brewing as Bobby Dazzler's Ale House in Perth in 2000. The brewery was closed for two years while Stallwood was recovering from injuries suffered while breaking up a fight, but resumed operations in 2006. In 2011, Nail earned global notoriety for brewing what was touted as "the most expensive bottle of beer in the world." Antarctic Nail Ale was made with water melted from a block of Antarctic ice. Of the 30 bottles that were produced, one was old at a Sydney charity auction for A$1,850. (At the time, Australian dollars were roughly in parity with US dollars.) More accessible products include 4.7 percent ABV Nail Ale and 10.6 percent Clout Stout.

As evidenced by the line of companies from Matilda Bay to Gage to Nail, Western Australia has emerged as an important brewing center. A checklist of other craft brewers in Western Australia includes the Bootleg Brewery, founded in 1994; Matso's Broome Brewery, founded in 1997; Colonial Brewing Company, founded in 2004; and Feral Brewing Company, founded in 2005.

Elsewhere in Australia, the beer traveler will find Burleigh Brewing in Queensland, founded in 2006; Copper Coast in South Australia, founded in 2005; Holgate Brewhouse in Victoria, founded in 2002; Lobethal Bierhaus in South Australia, founded in 2007; Moo Brew in Hobart, Tasmania, founded in 2005; and Mountain Goat Beer in Victoria, founded in 1996. Sydney craft brewers include St. Arnou, founded in 2001, and Skinny Blonde, founded in 2009.

While Australia has boasted some large international brewing companies in its brewing history, New Zealand greatly outnumbers its larger neighbor in the number of microbreweries and brewpubs. Like Australia, New Zealand experienced an industry consolidation in the latter twentieth century which resulted in all of its old established brewing companies being combined into three holding companies. These are Kirin-owned Lion Nathan (known in New Zealand as Lion Breweries) and BD Breweries, which are owned by Singapore-based Asia Pacific, as well as Independent Breweries, which are not independent but owned by Asahi in Japan.

Lion Nathan grew out of New Zealand Breweries of Aukland, which was founded as a ten-brewery consortium in 1923, with some components dating back to the 1840s. By 1977, when the company was renamed Lion Breweries, it was the largest brewer in New Zealand. In addition to its Lion Brewery in Aukland, the company owns and operates Speight's and Emerson's breweries, both based in Dunedin and founded in 1876 and 1993, respectively, as well as McCashin's Brewery (aka Mac's), a microbrewery started in Nelson in 1981. Their Canterbury Brewery was damaged beyond repair in the 2011 Christchurch Earthquake.

Above: Victoria Bitter is one of the most popular of the classic Carlton & United (now SABMiller) brands in Australia. (SABMiller)

Below: Mt. Macedon Ale is a 4.5 percent ABV pale ale brewed at the Holgate Brewhouse in Woodend, Victoria. (Holgate)

This page: These scenes from the Gage Roads Brewery include their Perth Beer Show award for Sleeping Giant IPA; general manager Aaron Heary with national sales manager Donald Pleasance; and founder/brewmaster Bill Hoedemaker adding hop pellets to a brew. (Gage Roads)

Steinlager is Lion's global flagship brand, a lager marketed in several variants including Steinlager Classic and Steinlager Pure, each of them in the range of 5 percent ABV, as well as 3.5 percent Steinlager Light. Originally called Steinecker, it was, according to the company, created in 1958 "in response to then Minister of Finance, Arnold Nordmeyer, who cut beer imports as part of his infamous Black Budget and challenged New Zealand brewers to compete by producing a lager of international quality." Renamed Steinlager in 1962 because the earlier name sounded too much like "Heineken," the beer was sold in a brown bottle in the domestic market and not put in green glass until it was first exported to North America in 1973. The green bottle was not used at home until 1977. The brand is used in

the parent company's sponsorship of yacht racing, which included the 1995 and 2000 wins by Team New Zealand of the America's Cup.

Other Lion brands include a family of 4 percent ABV lagers and ales marketed under the Speight's and Canterbury brand names as well as the McCashin craft brands. All top-fermented, these include 4.5 percent ABV Stoke Amber Ale, 5.2 percent Stoke Bomber Oatmeal Stout, and 4.5 percent Stoke Dark Porter.

Independent Breweries center on the Boundary Road Brewery in Papakura, a craft brewery started in 1987 by the late Michael Erceg and acquired by Asahi in 2011. In addition to its own brands, the Papakura facility exercises licenses obtained by Asahi to brew global brands such as Carlsberg, Tuborg, Holsten, and Kingfisher for the New Zealand market. The

Boundary Road brands include Spike's IPA and Spike's Red Rye Ale, both 6 percent ABV, as well as 4.2 percent Flying Fortress Pale Ale and 4.2 percent Bouncing Czech Pilsner.

DB Breweries dates back to 1930, when Sir Henry Kelliher and W. Joseph Coutts bought the Waitemata Brewery in Otahuhu, near Aukland. Other breweries still part of the group include the Tui Brewery in Mangatainoka, the Mainland Brewery in Timaru, and Monteith's Brewery in Greymouth. Long one of the most popular DB brands is a 4 percent ABV lager called Tui, which originated with Henry Wagstaff and Edward Russell, who started the Tui Brewery in 1889. The brewery was extensively remodeled in 2012. Products include 6.6 percent ABV Monteith's Barrel Aged Porter, 5 percent Monteith's Pilsener, 4 percent Monteith's Original Ale, and 7 percent Monteith's Doppelbock.

Still operating, the first brewpub in New Zealand was the Shakespeare Tavern, started in Aukland's Shakespeare Hotel in 1986 by Peter Barraclough and Barry Newman. Today, it is one of more than 70 microbreweries and brewpubs active in New Zealand. Another independent craft brewery is Josh Scott's Moa Brewing, started in 1999 and

named for New Zealand's enormous, albeit extinct, flightless bird. Moa brews a range of beers and is notable for its 10.2 percent ABV, barrel-aged Moa Imperial Stout.

Breweries in Australia and New Zealand tend to dominate the market across the Pacific south of Hawaii, but there are indigenous exceptions. Red Rooster Brewing in Koror, Palau produces ale and stout, both at 5 percent ABV, and operates a brewpub. Brasserie de Tahiti has been brewing in Papeete, French Polynesia since 1955. Their Hinano brand, a 5 percent lager, is distributed as far as Hawaii. South Pacific Brewing was started in Port Moresby in Papua New Guinea in 1952, and was later acquired by Heineken's Asia Pacific Breweries group.

Above: Serving Moa to fans attending the 2013 America's Cup. New Zealand's yacht, *Aotearoa*, lost the 19-race series. (Bill Yenne photo)

Below: Steinlager is New Zealand's best known brand. (Lion Nathan)

# Latin America and the Caribbean

Latin America, notably Mexico and Brazil, has the fastest-growing brewing industry in the world outside China. Production levels doubled in both countries during the 1970s, and they doubled again between 1980 and the mid-1990s. In 1980, Mexico and Brazil ranked eighth and fifteenth in the world, respectively, but since the turn of the century, Brazil edged out Germany for the number four spot behind China, the United States, and Russia, with Mexico in sixth. Brazil's total annual output rose from 12 million hectoliters in 1982, to 53 million in 1995, to more than 100 million by 2010. For Mexico, the respective figures are 20 million, 42 million, and more than 80 million. In the twenty-first century, with the United States just across the border, Mexico displaced the Netherlands (home of Heineken) as the world's largest beer exporting nation. Much of Heineken's production, it should be added, takes place outside the Netherlands.

Three Latin American lager brands have been among the top ten global beer brands since before the turn of the century. These are Brahma and Skol, both from Brazil, and Corona from Mexico. The latter is a product of Grupo Modelo, which has a two-thirds market share in Mexico.

The largest brewing conglomerate in the world, AB InBev, evolved directly from the Brazilian company

**Below: Enjoying Club Colombia in a nightclub in Bogota. This brand is a locally popular premium lager produced by Cervecería Bavaria. (SABMiller/Jason Alden/One Red Eye)**

AmBev, which was formed in 1999 as a merger of Brazil's two largest brewers. With the 2013 acquisition of Modelo, AB InBev controls the largest brands in Latin America as well as the Budweiser brands, the largest in North America.

As in much of the world outside North America and Western Europe, most of the major brewing companies in the Western Hemisphere are also involved in other food product lines, especially soft drinks and packaged fruit juices.

While most of what is brewed in Latin America today is pale lager, the Americas had a brewing tradition that predated the arrival of European beer. An indigenous native "black" beer was brewed in Brazil before 1492 and is still found in remote corners of the Upper Amazon. In the 1980s, the beer was "discovered" and nurtured for export by Alan Eames, the noted beer anthropologist. In the 1990s, a variant named Xingu, after a tributary of the Amazon, was brewed in export quantities at Cacador

Above: Relaxing with a bottle of Pilsen Trujillo at a Peruvian sidewalk cafe. The 5 percent ABV lager is part of the portfolio of Union de Cervecerías Peruanas Backus y Johnston, the leading brewing company in Peru. (SABMiller/Jason Alden/One Red Eye)

by Cervejaria Cacador. Today, it is brewed as a 4.7 percent ABV black lager by Cervejaria Sul Brasileira in Santa Maria, Brazil.

Of the many indigenous beer styles still brewed in Central and South America, one of the most popular is chicha, typically based on corn or cassava, but which can

**Above: Negra Modelo, a Munich-style dunkel lager, is perhaps the most widely exported Mexican dark beer. (Modelo)**

**Below: Corona Extra from Modelo is Mexico's best-selling beer, both domestically and in the export market. (Modelo)**

also be made from fruit. It was made during the halcyon days of the Inca Empire, which lasted until the sixteenth century, and it continues to be made and enjoyed in the twenty-first century.

The first Spanish breweries were established in the sixteenth century, but in the ensuing three centuries Latin American tastes paralleled those of Spain, just as North American tastes paralleled those of England. This meant an inclination toward wine and distilled spirits, such as mescal and tequila, rather than beer. There was also pulque, a fermented beverage favored in the nineteenth century by peasants. The climate of much of Latin America prevented the brewing of lager until the latter part of the century when ice-making technology became widely available. Early breweries were relatively small and brewed sencilla or corriente beer, which was similar to lager but fermented for a shorter period of time.

The first European-style brewery in the New World was probably the one started in Mexico City by Don Alonso de Herrera that began brewing in 1544. There was an influx of German immigrants in the mid-nineteenth century and this opened the door for lager brewing. Cervecería Toluca in

Toluca, 20 miles from Mexico City, was started in 1865 by Augustin Marendaz, a Swiss immigrant. In 1875, it was acquired by Santiago Graf, who imported one of the first ice-making machines into Mexico in 1882. His new ice machines helped him branch into lager brewing, and Graf became one of Mexico's largest and most successful brewers.

Throughout the twentieth century, Mexico's brewing industry was traditionally dominated by a Big Three, Cuauhtémoc, Moctezuma, and Modelo, although the former two were controlled by members of the same families and owned under the same holding company. Growth during the early twentieth century was aided considerably by Prohibition in the United States, during which beer drinkers traveled south, and the beer traveled north. The Big Three also grew though acquisition of smaller breweries across Mexico.

Cervecería Modelo, the largest of the Big Three, evolved from Santiago Graf's Cervecería Toluca, taking on the Modelo name in 1925, as the center of brewing operations was concentrated at a new brewery in Mexico City. The company expanded through acquisitions, such as that of Cervecería de Pacifico in Mazatlan, famous for the Pacifico brand, in 1954. Pacifico was established in 1900 by Jacob Schuehle, who had also built the original Moctezuma brewery.

In the cases of acquisitions, the acquired brands were retained.

In 2013, after several years of on-and-off corporate courtship, Modelo became part of AB InBev.

Modelo's flagship brand is Corona, the largest brand in Mexico, the largest Mexican brand sold elsewhere in the world, and a long-time member of the global top ten brands. In the 1980s, when the popularity of Mexican cuisine and the proliferation of Mexican-themed restaurants in the United States enjoyed a steep increase, sales of Mexican beers also rose, and those of Corona skyrocketed. Mexico surpassed Canada as North America's second-biggest brewing nation. In 1986 alone, Corona increased its share by an incredible 169 percent, making it second only to Heineken among imported beers in the United States market. In 1997, it overtook the Dutch brand and has maintained this position since.

A 4.6 percent ABV, flavor-neutral lager made with a sizable proportion of inexpensive non-barley grain, Corona earns consistently low marks from beer critics, but there is no denying its immense popularity. Higher ratings are given to Modelo's 5.5 percent dark lager, Negra Modelo. First introduced in Mexico in 1930, it has its roots in the dark lager tradition which existed in Mexico in the nineteenth century, and it is offered as a choice at most Mexican restaurants in the

United States. Modelo also markets 4.4 percent ABV Modelo Especial, first introduced in 1925, and 4.5 percent ABV Pacifico Clara, a very light lager.

Cervecería Cuauhtémoc in Monterrey, meanwhile, was the first big lager brewery built on the large scale of the American and European breweries of the time. It was founded in 1890 by related members of the Calderon, Garza, Muguerza, and Sada families, together with Wilhelm Hasse and Joseph Schneider.

Cervecería Moctezuma was established in 1894 at Orizaba, about 100 miles southeast of Mexico City under the name Cervecería Guillermo Hasse. In this venture, Wilhelm (Guillermo in Spanish) Hasse was joined by Cuno van Alten, Adolph Burhard, and Henry Manthey. With investments from the Calderon, Garza, Muguerza, and

**Above:** Classic labels from two of Cervecería Moctezuma's best-known brands. There are several variations on Dos Equis (XX), including a pale lager and an amber Vienna-style lager. Variations on Sol include several pale lagers, low alcohol lagers, and lagers infused with lemon and other flavorings. (These images from the author's collection were supplied by request by the breweries in the 1990s)

**Below:** The lager product line from Cervecería Cuauhtémoc as seen before the brewery officially merged its operations with Moctezuma. Except for Brisa, the other brands, Bohemia, Carta Blanca and Tecate, remain in production, with substantially unchanged labels. (Author's collection)

This page: Images of Mario Garcia and his Cervecería Cucapá in Baja California. At the right, he leads a brewery tour. Below are his Cucapá Clasica and Chupacabras Pale Ale, the latter named for the mysterious nocturnal creature. (Cucapá)

Sada families, the brewery, now called Moctezuma, came under a new family-owned holding company, Valores Industriales, which was created in 1936. As with rival Grupo Modelo, the Valores group grew through acquisitions during the twentieth century. One of its notable purchases was Tecate in the Baja California city of the same name. Tecate began brewing in 1943 and was acquired in 1955. As with most Big Three acquisitions, the brand identity was retained and exists today as a 4.5 percent ABV lager.

Cuauhtémoc and Moctezuma continued to operate as autonomous entities within Valores until 1986, when they were combined into an entity appropriately named Cuauhtémoc Moctezuma. In turn, this brewing entity was folded into the Valores beverage holding company, Fomento Económico Mexicano, SA (FEMSA), which

also contained what was described variously as the largest or second-largest bottler of Coca-Cola in the world. In 2010, Heineken acquired FEMSA's brands and breweries in exchange for a 20 percent share of Heineken stock. With brewing continuing at Guadalajara, Monterrey, Navojoa, Orizaba, Tecate, and Toluca, the traditional Cuauhtémoc and Moctezuma brand names continue to be used.

Since it was introduced in 1905, the flagship lager from Cuauhtémoc has been Bohemia, of which several similar products are marketed, ranging from the 4.8 percent ABV Bohemia Clasica to the slightly darker Bohemia Obscura. Another popular member of the family is Carta Blanca, a 4.5 percent lager launched in 1890.

Moctezuma's flagship brand is Dos Equis (XX), a lager first brewed by Wilhelm Hasse in 1897. It was named for the impending

twentieth century, with XX being the Roman numeral for 20. The two principal variants are the light-colored 4.4 percent ABV Dos Equis Special and the 4.7 percent Dos Equis Amber.

Since the turn of the twenty-first century, Mexico has become home to a long overdue, albeit small to date, wave of craft breweries. Worthy of note is Cervecería Cucapá (formerly Cervecería de Baja California), founded in 2002 in Mexicali by Mario Garcia. Originally a brewpub, the brewery has been bottling their Cucapá brand, named for one of the indigenous tribes who once lived in the lower Rio Grande Valley, since 2006. The beer is now being exported to California and the American Southwest. As the brewery notes, "We are the first and only Mexican micro-brewery that has received a score over 90 from the Beverage Tasting Institute for four of our beers. Our beer styles range dramatically from a Blond Ale to a Barley wine. Our beers are made with local ingredients, which give them that Mexican touch."

Brands include 4.5 percent ABV Cucapá Clasica, 4.5 percent Cucapá Jefe (a wheat ale), 4.5 percent Cucapá Honey, 4.5 percent Cucapá Obscura, 7.5 percent Cucapá Runaway IPA, 10 percent Cucapá Tequila Barrel Aged Barleywine, and 5.8 percent Cucapá Chupacabras Pale Ale, named for the popular and mysterious, and probably mythical, nocturnal creature that is the Mexican analog of the North American bigfoot.

Central America has a long brewing tradition, where German style lagers and pilsners are much more common than in Mexico, owing to the region's more tropical climate.

Cervecería Hondurena in Honduras evolved through mergers with different companies that operated generally between 1915 and 1965. These companies individually marketed lager brands such as Salva Vida, born in 1916 in La Ceiba; Imperial, first brewed in 1930; Nacional, born in Tegucigalpa in 1953, and the Ulva brand, first brewed in San Pedro Sula in 1928. The products of the brewery today include 4.8 percent ABV Port Royal and 5 percent Imperial. The Port Royal and Salva Vida brands were exported to the United States beginning in 1985 and 1988, respectively.

Cervecería Nacional of Panama City was founded in 1909 and is Panama's largest brewer. The company's lager brands include 3.8 percent ABV Atlas and 4.8 percent Balboa Premium. The company owns Cervecería Chiricana and is itself owned by Grupo Bavaria of Colombia, a SABMiller company.

In Guatemala, Cervecería Centro Americana dates back to 1886 and has brewed its 4.4 percent ABV Gallo lager, albeit at varying ABV ratings, since 1896.

This page: Imperial lager is one of the lead products from Cervecería Hondurena, the principal brewing company in Honduras. Other brands, all lagers, include Port Royal, Barena, and Salva Vida. The brewery is in San Pedro Sula. (SABMiller/Jason Alden/One Red Eye)

Above: Pouring a glass of Port Royal lager at a trendy Honduran night spot. (SABMiller/Jason Alden/One Red Eye)

Below: Tona lager from Nicaragua's Industrial Cervecería and four from Guatemala's Cervecería Centro Americana: Victoria, Moza Obscura (Dark) bock, Moza Gold bock, and Famosa (Gallo) lager. (US Brands (Tona)/ Centro Americana)

Another of the company's popular lagers is 5 percent Victoria.

In Nicaragua, Cervecería de Nicaragua established a reputation with its Cerveza Xolotlan, introduced until 1929. The company spent most of the twentieth century vying with rival Industrial Cervecería SA (ICSA) for market share before the two firms

merged in 1996. To the marriage, the former company brought its 4.5 percent ABV Victoria brand, first introduced in 1942, while ICSA brought its leading lager, 4.6 percent ABV Tona, which it had introduced in 1977 to compete with Victoria. Tona was first exported to the United States in 2003.

In 1908, the brothers Rupert Cecilio Morales and Stanley Lindo Morales relocated from Jamaica to Costa Rica and started a business called the Florida Ice & Farm Company (FIFCO). They sold ice and became involved in the produce business. Over time, the company became very big in soft drink bottling and resort hotels. In 1912, they bought a brewery, and in 1958, they bought their biggest competitor in the beer business and rolled the two together in

1966 as Cervecería Costa Rica, which became famous for its Imperial lager. In 2012, the company bought a basket of breweries from KPS Capital Partners of New York. The basket was called North American Breweries Holdings, and in the basket were two notable United States brewing companies, Genesee, a long-established regional brewer from upstate New York, and Magic Hat, a craft brewer from Vermont, which had just acquired Seattle-based Pyramid Breweries. Also in the basket were the United States operations of Labatt, one of Canada's two largest brewers.

The beer traveler will find that the brewing tradition of the little golden flecks which are the islands of Caribbean is long and varied. By the seventeenth century, as brewing was taking hold on the mainland to the north, the isles of the West Indies had already gone over en masse to the warm embrace of rum. At the beginning of the nineteenth century, Guinness was exporting its West Indies Porter, the precursor to Guinness Foreign Extra Stout.

By 1898, there were just seven breweries in the entire region: four in Jamaica and one each in Barbados, Trinidad, and Cuba. When the United States defeated Spain in the Spanish American War, Obermeyer & Liebmann came south from Brooklyn to open a brewery in Cuba's capital.

During the early twentieth century, the leading brand in Cuba was Hatuey, first introduced in 1927 and brewed in Santiago by the large Cuba-based rum distilling company Bacardi. A new brewery, designed by the architect Enrique Luis Varela, was built in Havana in 1948. Hatuey became so integral to the Cuban experience that Ernest Hemingway mentioned it in his novel *The Old Man and the Sea*. When Hemingway was awarded the 1954 Nobel Prize in Literature, in part because of this book, they threw him a party at the brewery, which was a short walk from his home in Havana. In 1960, when Fidel Castro took over in Cuba, his government confiscated everything owned by Bacardi, and Hatuey faded from the scene. In 2012, Cuban immigrant Anler Morejon revived the brand, brewing it for distribution in Florida from the Thomas Creek Brewery in Greenville, South Carolina.

Today, the most popular beer in Cuba, with a reported 80 percent market share, is Cerveza Cristal. A 4.9 percent ABV lager, Cristal is brewed by Cervecería Bucanero in Holguin, which is a joint venture between Labatt of Canada and the Cuban government. The brewery also produces 5.4 percent ABV Bucanero Fuerte and 6.5 percent Bucanero Max.

In Santo Domingo in the Dominican Republic in 1929, American brewer Charles Wanzer founded Cervecería Nacional

**Above:** During the early twentieth century, Hatuey was the signature beer of Cuba. Even Hemingway made reference to it. (Author's collection)

**Below:** Red Stripe from Desnoes & Geddes is the signature lager of Jamaica, and for a time, it was that of the culture surrounding the Reggae musical experience. (Diageo)

212

Above: Relishing a glass of Club Colombia. (SABMiller)

Below: Lagers from Cervecería Bavaria include 4.7 percent ABV Club Colombia, the flagship 4 percent Aguila, 4 percent Poker and 5 percent Pilsen Night. The Bavaria press release accompanying the 2012 release of the latter notes that it is "the new protagonist of the evening and the rumba. . . [arriving] in the market under the concept of the 'lo inesperado de la noche,' referring to the way the youth segment of Pilsen lives the rumba, as the best nights are those which are not planned." (SABMiller)

Dominicana, which grew into the largest brewer in Central America and the Caribbean. Today, it is owned jointly by AB InBev and Grupo Leon Jimenes, a holding company that owns La Aurora, makers of Aurora and Leon Jimenes cigars. Dominicana also brews Heineken and SABMiller products under license. The flagship among Dominicana's own products is Presidente, a 5 percent ABV lager first marketed in 1935.

Desnoes & Geddes of Kingston, Jamaica, was founded in 1918 as a soft drink business by Eugene Desnoes and Thomas Geddes. During the twentieth century, they became famous for their Red Stripe lager, which became a popular export to the United Kingdom and the United States

in the 1970s as Jamaican Reggae music also developed a huge following abroad. Also in the 1970s, Guinness approached Desnoes & Geddes about a possible contract brewing arrangement for Guinness Foreign Extra Stout. This didn't work out, so Guinness decided to build a brewery on the island and formed Guinness Jamaica Limited. This facility opened in 1973 against the backdrop of civil unrest and an economic downturn that followed the election of a Marxist government. As Guinness Jamaica Limited struggled, Desnoes & Geddes cut a deal to contract brew for Heineken in Jamaica. In 1985, Guinness sold its Jamaica subsidiary to Desnoes & Geddes, who continued to brew Foreign Extra Stout under license. In 1993, Guinness PLC bought a

controlling interest in Desnoes & Geddes. Today, a family of Red Stripes is brewed, ranging from 3.6 percent ABV Red Stripe Light to 4.7 percent Red Stripe Jamaican Lager to 6 percent Red Stripe Bold. Desnoes & Geddes brews its own 10 percent ABV Dragon Stout as well as Foreign Extra Stout for the parent company.

As the beer traveler moves south from Central America and the Caribbean, a first stop might be in Colombia. Here, the most commonly seen brands are those of Cervecería Bavaria. These were acquired as Bavaria gradually bought other Colombian brewing companies during the mid-twentieth century. The company was originally founded in 1889 by the German brewer Leo Kopp as the Deutsche Bräuerei Bavaria. Important early competitors that were later acquired by Bavaria included Cervecería Unión, founded in 1902, and Cervecería Barranquilla, founded in 1913 and soon famous for its Aguila (Eagle) brand. So important was the brand that in 1967, when Cervecería Barranquilla merged with Cervecería Bolivar, they renamed themselves Cervecería Aguila. Later the same year, Bavaria acquired Cervecería Aguila, but kept the name for the subsidiary and continued using the brand name. Cervecería Unión was added to the Bavaria family in 1972.

Bavaria went public on the Colombia Stock Exchange in 1981, and with further acquisitions and expansion, renamed itself Grupo Empresarial Bavaria (GEB) in 1996, with the brewing sector known simply as Bavaria. A round of international acquisitions put Bavaria on the world stage. Central de Cervejas in Portugal was acquired in 1990, and in 2001 GEB purchased a 91.5 percent stake in Panama's Cervecería Nacional. The following year, GEB bought a controlling interest in Backus y Johnston, which controls virtually the entire beer market in Peru. By 2004, GEB was the tenth-largest brewing group in the world. A year later, they became a part of SABMiller.

Leading brands from Bavaria, all Pilsners, include 4 percent ABV Aguila, 6.5 percent Brava, 4.7 percent Club Colombia, 4 percent Costeña, 4.2 percent Pilsen, and 4 percent Poker.

Other smaller Colombian brewing companies include Bogotá Beer Company (BBC), Cervecería

**Above: Pouring a glass of Aguila, Colombia's favorite beer brand.** (SABMiller/Jason Alden/One Red Eye)

**Below: The four finalists in the Chicas Aguila 2012 pageant were Laura Victoria Gamboa, Johanna Milena Díaz, Carolina Landaeta, and Melissa Porras. The quartet of Chicas travel the country, promoting the product at cultural events, festivals, and public fairs.** (Cervecería Bavaria)

This page: Cusqueña lager was first brewed in Cusco, Peru by Cervecería Cusqueña in 1911. (SABMiller/Phil Meech/One Red Eye)

Colón, Cervecería San Tomas, and Cervecería Tres Cordilleras.

As noted above, Backus y Johnston is Peru's largest brewing and beverage company. It was founded in Lima in 1879 by the English brewers Jacob Backus and John Howard Johnston, who organized it as a British company in order to raise capital on the London Stock Exchange. Though both founders were no longer involved by the turn of the twentieth century, the brewery remained a British company until 1954.

Now officially known as Union de Cervecerías Peruanas Backus y Johnston, the company's growth to its present market prominence came through the acquisition of smaller breweries, especially the takeover in 1994 of its biggest rival, Compañía Nacional de Cerveza, which dated back to 1863 and whose Pilsen Callao brand was well known in South America and on the shelves of specialty shops in North America. Other acquisitions have included the Cervecería Arequipeña in Arequipa, Cervecería Cusqueña in Cusco, and Cervecería Pucallpa in Pucallpa. Backus y Johnston also operates breweries in Chiclayo and Trujillo, and the flagship brewery is still in Lima. The product line includes 5.2 percent ABV Pilsen Callao and 5 percent Pilsen Trujillo as well as Cristal, a 5 percent light lager and Malta Pilsen Polar, a 5.5 percent ABV dark lager.

Chile and Argentina have shared similar brewing traditions and a major brewery. Here, as in most of Latin America, traditional brewing of chicha coexists with modern industrial brewing that originated in the nineteenth century with German immigrants. Meanwhile, both countries have well-established wine industries whose products are highly regarded and exported throughout the world. Despite Argentinean and Chilean wine getting a great deal more international attention than their beer, the latter is quite popular at home.

The largest brewing company in Chile is Compañía de las Cervecerías Unidas (CCU), which dates back to a brewery founded in 1902 and is an amalgam of many other smaller brewing companies. Its subsidiary Compañía de Cervecerías Unidas Argentina (CCUA), started in 1995 and now partly owned by AB InBev, is one of the top two or three brewers in Argentina. In Chile, the leading brands are Cerveza Cristal, Escudo, and Dorado, although CCU had a license from Heineken to brew its flagship lager. For CCUA, the flagship brand is Schneider, a 5 percent ABV lager. CCUA, meanwhile has contracts to brew both AB InBev and Guinness products.

In Chile, meanwhile, AB InBev has an interest in Cervecería Chile, which was started in 1995 and acquired in part in 2003 by InBev predecessor AmBev. The current product line includes local brands such as Becker, Quilmes, and Pacena as well as parent company international brands, including Beck's, Brahma and Stella Artois.

Also part of AB InBev's Argentina portfolio since 2006, the products of Cervecería Quilmes dominate the Argentine market with a three-quarters share. Founded by German brewer Otto Bemberg in 1888 in Quilmes, near Buenos Aires, the brewery went on to make its brand a virtual national symbol in the twentieth century. Today, while operating breweries in five Argentine cities, Quilmes is widely exported across the continent and the world. The flagship products, Quilmes and Andes, are both 4.9 percent ABV lagers, but there is also a 5.6 percent Andes Porter.

Of perhaps greater interest to the beer traveler, rather than mass market lagers from internationally affiliated large brewers, are the offerings of a new generation of craft brewers. Notable among these is Cervecería Kunstmann in Valdivia, Chile. Started in 1997 by the German-Chilean Kunstmann family, it is one of the first breweries in southern Chile since the Anwandter Brewery was destroyed in the 1960 earthquake. Kunstmann

## Oktoberfests in Latin America?

The largest beer festival in South America is the **Oktoberfest** held annually for 18 days in October in the Parquet Vila Germanic (Germanic Village Park) in the Brazilian city of Blumenauer. Averaging more than 600,000 attendees, it is considered to be one of the two or three largest **Oktoberfest** celebrations in the world. Other Brazilian **Oktoberfests** include those in Santa Cruz do Sul and Igrejinha, which host around 500,000 and 200,00 visitors annually. If the beer traveler who visits these cities in October decides to stay on in November, the **Mondial de la Bière**, based on the one held annually in Montreal, is held for three days in the middle of the month at the Espaço Ação Cidadania in Rio De Janeiro.

Meanwhile, Argentina's largest **Oktoberfest** is in Villa General Belgrano in the Córdoba Province, while Kunstmann hosts its own **Bierfest** in Valdivia. Farther north, Mexico's German heritage is revealed in annual **Oktoberfests** in Puebla, Mexico City, Mazatlán and Tapachula.

exports its products throughout the world, while also maintaining a brewery restaurant in a pleasant rural setting near the city. Kunstmann produces two pale ales, 5 percent ABV Torobayo and 7.5 percent Gran Torobayo, as well as 5 percent Valdivia Blonde, 4.8 percent Honig Ale, and 5.3 percent Bock.

**Above: Compañía Cervecería Kunstmann SA in Valdivia, Chile is one of many smaller German-style breweries in South America which were founded by German immigrants and their descendants. The Kunstmann Oktoberfest is one of the great German-Chilean festivals held annually in the country. (Kunstmann)**

Above: Pilsen Callao is a leading Peruvian lager. (SABMiller/Jason Alden/One Red Eye)

Below: AmBev (later AB InBev) evolved out of Brazil's Companhia Cervejaria Brahma, and the Brahma brand has long been in the global top ten. Other AB InBev Brazilian brands include Antarctica, Cristal, Bohemia Escura (Dark), and Bohemia Pilsen. Antarctica and Bohemia are brewed at Companhia Brasileira de Bebidas. (AB InBev)

Started in 2003, Kross Microbrew Beer (aka Southern Brewing) in Curacaví, Chile, has earned awards for a product line that includes 5 percent ABV Kross Golden Ale, 5.5 percent Kross Lupulus Ale, and 6.5 percent Kross Maibock. Then there are the interesting regional breweries. An example is Cervecería Austral in Punta Arenas, overlooking the Strait of Magellan in far southern Chile, which was founded in 1896 as Cervecería Patagonia. Its products include 4.6 percent ABV Austral Lager, 4 percent Austral Pale Ale, and 5 percent Yagan Dark Ale.

Brazil, South America's largest economy and largest beer market, is the backyard of Bebidas das Américas (AmBev), the heart and predecessor of global world leader AB InBev. Two of their biggest Brazilian brands, Antarctica and Brahma, are among the oldest in Brazil, with their Bohemia brand, dating to 1853, being the oldest. Cervejaria Brahma originated in 1888 and became the national leader during the twentieth century with its Brahma Chopp brand. In 1999, AmBev was formed through the merger of Brahma and its biggest rival, Antarctica Paulista.

Brahma flagship products include 4.6 percent ABV Brahma, 5 percent Brahma Chopp, and 4 percent Brahma Malzbier, a black lager. Other beers are 5 per-

cent Antarctica Pilsen, 5 percent Bohemia, 5 percent Bohemia Escura (a black lager), and a 5.6 percent hefeweizen called Bohemia Weiss. AB InBev also owns the South America rights to Skol, a multinational lager brand originally created in 1959 and now controlled by Carlsberg in most of the world. Though forgettable for its flavor, Skol has been one of the world's half-dozen best-selling beer brands for many years.

In turn, the AmBev merger elevated Cervejarias Kaiser, founded in 1982, to the short-lived status of Brazil's second-largest brewing company, although its market share was around 15 percent. Between 2002 and 2006, Kaiser was owned by Molson, after which controlling interest was sold to Mexico's FEMSA, which became part of Heineken in 2010. Products include lagers marketed under the Kaiser and Bavaria names, as well as pilsners such as 4.5 percent ABV Bavaria and 5 percent Bavaria Pilsner.

Succeeding Kaiser as Brazil's number two brewer, with a market share of around 10 percent, is Brasil Kirin. Founded as Cervejarias Schincariol in 1939 in Itu, near Sao Paulo, the company and its 13 breweries was acquired in 2011 by the Kirin Brewery Company of Japan. The traditional core lager brands include 4.4 percent Glacial, 4.7

percent Nova Schin Pilsen, and 4.2 percent Nova Schin Munich, a dark lager. A unique part of Brasil Kirin's business model has been the acquisition of craft breweries which have appeared in Brazil since the beginning of the twenty-first century. Among these was Baden Baden of Campos do Jordao, which had been recognized as Brazil's first microbrewery.

A growing independent Brazilian craft brewer is the German-themed Cervejaria Sudbrack, which brews under the Eisenbahn brand name in Blumenau. Lagers include 4.8 percent ABV Pilsen and 4.8 percent Dunkel. Ales include 4.8 percent Weizenbier, 8 percent Weizenbock, and 6.3 percent Weinachts Ale for Christmas.

**Above: A Peruvian Paso horseman pours a glass of Cristal from Backus y Johnston. (SABMiller/Jason Alden/ One Red Eye)**

**Below: Eisenbahn Weinachts Ale, the Christmas beer brewed by Cervejaria Sudbrack in Blumenau, Brazil. (Sudbrack)**

# Canada

Canada is the fourth-largest brewing nation in the Western Hemisphere behind the United States, Brazil, and Mexico, having relinquished third place to Mexico in the 1970s. It has maintained a constant annual output of around 20 million hectoliters ever since, while its neighbors experienced varying degrees of growth.

Through the twentieth century, the Canadian market was dominated by a Big Three of national brewing companies—John Labatt, Limited (Brasserie Labatt in Quebec), Molson Breweries, Limited (La Brasserie Molson), and Carling O'Keefe—plus the much smaller Moosehead Breweries in the Maritimes. However, by the beginning of the twenty-first century, the Big Three had disappeared into conglomerates, Moosehead was Canada's largest Canadian-owned brewer, and the news in Canadian brewing was all about the maturing craft brewing scene.

The leading brewers today, as calculated on the basis of the best beers in Canada as rated by the readers of *Beer Advocate*, are Brasserie Dieu Du Ciel in Montréal; Central City Brewing in Surrey, British Columbia; Driftwood Brewery in Victoria, British Columbia; Les Trois Mousquetaires in Brossard, Quebec; Microbrasserie Charlevoix in Baie-Saint-Paul, Quebec; and Unibroue in Chambly, Quebec.

Brewing has been a part of the lives of Canadians since the seventeenth century, with the first commercial brewing company said

**Below: The Molson brewery, with the Montreal skyline in the background. John Molson started the company on this same site in 1786, and some of the same cellars that were used by the company in the beginning are still in use today. Molson is North America's oldest brewing company, and the second oldest company of any kind, after the Hudson's Bay Company, in Canada. (Molson)**

to be that of Jean Talon, which opened in Quebec in 1668. John Molson arrived from England in 1782 and started his Montréal brewery, now the oldest in North America, in 1786. During the nineteenth century, all of the familiar names which would dominate the rise of the big brewing empires of the twentieth century arrived to take their place alongside Molson. Alexander Keith started brewing in Halifax in 1820, Thomas Carling set up shop in London, Ontario, in 1840, and in 1847 John Labatt and Samuel Eccles acquired a brewery that was started in the same city in 1828 by innkeeper George Balkwill. In 1853, Labatt became the sole owner and renamed the company for himself. Eugene O'Keefe purchased the Hannath & Hart Brewery in 1862 and renamed it for himself in 1864. In 1867, John and Susannah Oland started brewing in their backyard in Dartmouth, across the harbor from Halifax, Nova Scotia.

During the early twentieth century, Prohibition destroyed all but the most hardy of United States brewing companies, but this did not happen to the same extent north of the border. The groundswell of support for a total ban on sales of alcoholic beverages that swept through North America culminated in Prohibition that lasted in the United States from 1920 to 1933. In Canada, meanwhile, prohibition was

**Above: Inside the brewhouse at the Central City Brewing Company in Surrey, British Columbia. (Central City)**

Above: Classic twentieth century labels from Canada's twentieth century "Big Three" brewing companies, Molson, Carling and Labatt. (Author's collection)

Below: Kokanee from the Labatt-owned Columbia Brewery in Creston, British Columbia, has been long been an iconic western Canadian brand. It is known as "the Beer Out Here." (Labatt)

implemented by province, rather than on a national level. In provinces other than Quebec, where prohibition lasted for a matter of weeks in 1919, prohibition was enacted in 1916–1917 and repealed in 1921 in British Columbia and Manitoba, 1924 in Alberta and Ontario, and by the end of the decade it was over in most of the country.

Around this time, the business tycoon Edward Plunket "E.P." Taylor was assembling a consortium of breweries that he called Canadian Breweries Limited. Between 1930 and 1934, Carling and O'Keefe were added to his collection, effectively creating the company that would later be known as Carling O'Keefe.

From this point through to almost the end of the century, it was the era of Canada's Big Three brewing companies, Labatt, Molson, and Carling O'Keefe. Brewing in Canada was a big boys' game with a high price of admission. Because of laws that required beer to be brewed in the province where it was sold, a national brew-

ing company needed to have a brewing facility in every major province. During their heyday, through most of the twentieth century, each of the Big Three operated around a dozen breweries across Canada, some built from scratch and others acquired though buying out existing companies. Inter-provincial trade barriers were finally relaxed in 1992.

In the meantime, there were opportunities for regional breweries to compete in specific provinces. In Nova Scotia, the story of the Keith and Oland breweries became intertwined, but George Oland, the son of John and Susannah, moved to New Brunswick in 1928 and took over Ready's Breweries in Saint John. A company milestone occurred in 1931, when George Oland rechristened his ale "Moosehead." In 1947, the company name was changed from New Brunswick Breweries to Moosehead Breweries, and Ready's Pale Ale was rechristened as "Moosehead Pale Ale" to mark the company's entry into the Nova Scotia market.

Above: Twentieth century labels from brands associated with eastern Canada include Oland from Halifax, Nova Scotia; Moosehead from St. John, New Brunswick; and O'Keefe, brewed in Etobicoke, Ontario. (Author's collection)

Below: Once confined to the Maritime provinces, Moosehead is now a national brand, and the company is Canada's largest independent brewery. (Moosehead)

In 1971, the Halifax branch of the Oland family, which ran a competitive operation called Keith's Brewery, decided to sell to Labatt. This left Moosehead Breweries, with plants in Saint John, New Brunswick, and Dartmouth, Nova Scotia, as the last major Canadian independent brewery. It was in 1978 that Moosehead entered the United States market with their Moosehead Canadian Lager beer in bottles, and in 1984 Moosehead draft beer was introduced. By the 1990s, Moosehead was available in all 50 states and was ranked as the seventh-largest import beer out of 450 brands. In 1985, Moosehead arrived in England, distributed by Whitbread. It was not until 1992 that Moosehead Breweries introduced its beer in the Canadian provinces of Ontario, British Columbia, Alberta, and Newfoundland.

Two prominent twentieth-century regional breweries in Ontario, Canada's largest province, were Northern Breweries of Sault Ste. Marie and Hamilton Breweries in Hamilton. Founded in 1981, Hamilton brewed Grizzly, a lager originally brewed only for the United States export market but later available in Ontario as well. The brewery was sold to Heineken and operated as Amstel Breweries Canada Limited before being sold in 1992 and renamed as Lakeport Brewing. In 2007, Labatt bought Lakeport and announced the closing of the brewery.

Sleeman Brewing was founded in Guelph, Ontario, by John Sleeman in 1834. John's son, George, managed the brewery until 1933 when, during the last year of Prohibition, he was caught smuggling beer into Michigan. In 1985, with backing from Stroh Brewing, his great-grandson, John Sleeman, once again incorporated the brewery. In 1988, Sleeman Cream Ale, the old flagship brand, was reintroduced.

In western Canada, several notable independent regional brands and breweries came and went, some of which survived into the twenty-first century. One

such brand is Kokanee lager, first brewed in 1959 in Creston, British Columbia, by the Columbia Brewery. The brewery was acquired by Labatt in 1974, but the brand still remains, and it is hard to travel anywhere in British Columbia or northwestern Montana and not see Kokanee on cooler shelves.

Another British Columbia brewery that has seen many twists and turns is the one near a fresh water spring in Prince George that is now known as Pacific Western. Founded in 1957 as Caribou Brewing, it was sold to Carling O'Keefe in 1962 and immediately auctioned off, going to maverick entrepreneur "Uncle Ben" Ginter, who named it Tartan Breweries. Around this same time, Ginter also bought Rocky Mountain Brewing of Red Deer, Alberta.

"No story about the Canadian beer business would be complete without Ben Ginter," writes Allen Winn Sneath in *Brewed in Canada*. "Although he was only a player for fifteen years, his unorthodox approach to the business reserves him a permanent place in Canada's brewing history." Ginter also put his beer into cans before the Big Three and was a pioneer in the use of the pull tab. In 1991, the Prince George brewery, now called Pacific Western, was taken over by Kazuko Komatsu, who would successfully guide it into the twenty-first century.

As the last decade of the twentieth century approached, winds of mergers and acquisitions brought down the curtain on the golden age of the Big Three. Of the Big Three, Labatt's had been

**Above:** Launched in 1951, Labatt Blue was Canada's largest-selling brand from 1979 until overtaken by Budweiser in 2004 (Labatt)

**Below:** Checking the brew inside a Labatt brewhouse. (Labatt)

**Right:** Products from Great Western Brewing in Saskatoon, Saskatchewan. (Great Western)

Canada's largest brewing company, while Molson traditionally had the largest Canadian share of the lucrative export market to the United States. In 1989, however, Molson became the biggest as it bought third-place Carling O'Keefe, owned by Australia's Elders IXL since 1987. Rights to the popular Carling Black Label brand name outside Canada were sold or licensed to various entities.

In 1995, Labatt was acquired by Interbrew, the predecessor to AB InBev, although Labatt products were sold in the United States by an autonomous entity known as LabattUSA. In 2009, the latter entity was divorced entirely from Labatt and sold to the investment firm of KPS Capital Partners, who formed a holding company called North American Breweries, an entity based on the Genesee Brewing Company in Rochester, New York. In turn, North American Breweries was sold in 2012 to Costa Rica–based Florida Ice & Farm, the parent company of Cervecería Costa Rica.

Labatt's flagship beer is Labatt Blue, a 5 percent ABV pale lager, which eclipsed Labatt 50, a 5 percent blonde ale as Canada's best-selling beer brand in 1979. Labatt 500 is still part of the portfolio. As with major American national brands, Labatt developed a low-calorie "light" lager brand and a "dry beer" as a line extension in the 1970s and 1980s, respectively. Some of Labatt's entries in these fields, all pale lagers, include Labatt Blue Light and Labatt Lite, both at 4 percent, and 5.5 percent Labatt Extra Dry.

In 2005, Molson merged with Coors in the United States to form the Molson Coors Brewing Company, the world's seventh-largest brewing company. Is it a Canadian company or an American company? The answer is hard to intuit. Ownership and management are shared equally by the Coors and Molson families. The company is "dual headquartered" in Denver and Montréal,

Above: Alexander Keith's IPA, brewed in Halifax, Nova Scotia for over a century. (Labatt)

Below: Classic labels of still-extant early Canadian craft brewers: Big Rock of Calgary; Brick of Waterloo, Ontario; and Upper Canada of Guelph, Ontario. (Author's collection)

Above: The eclectic product line of Nelson Brewing in Nelson, British Columbia. (Nelson)

Below: Current labels for two of Molson's oldest brands; and updated packaging for Carling lager. All are now part of the Molson Coors portfolio. (Molson Coors)

Coors), which was created with its own completely new logo and a separate headquarters in Chicago. The purpose of MillerCoors was to market the Molson, Miller, and Coors brands as well as those of Pabst, but not other brands owned by these companies, in the United States, but not elsewhere.

Molson's flagship brand is Molson Canadian, a 5 percent ABV pale lager introduced in 1959, which is accompanied at the head of the portfolio by the similar 5 percent pale lagers Molson Golden and Molson Export. The Carling legacy is represented by several Carling-branded products, all pale lagers, including the classic Carling Black Label at 5 percent ABV. The O'Keefe legacy is represented by 5 percent O'Keefe Ale, one of only a few non-lagers in the Molson lineup. There are "Light," "Dry," and "Ice" variants of the Carling and Molson brands, all pale lagers.

With Labatt and Molson going from a Big Two to components of international conglomerates, Moosehead emerged as Canada's oldest and largest independent brewery. In the meantime, the elimination of inter-provincial trade barriers had allowed the brand to become truly national. In 2008, the sixth generation of the founding family took the reins of the company as Andrew Oland became Moosehead's president.

Moosehead's longtime flagship brand, Moosehead Lager has 5 per-

and shares are traded on stock exchanges in both countries.

In 2007, further complicating the corporate web for the observer, Molson Coors entered a joint venture with SABMiller that was called MillerCoors (not Molson-Miller-

cent ABV. So too does a lager marketed under the James Ready name, an homage to Ready's Breweries in Saint John, absorbed by Moosehead in 1928. There is also a James Ready 5.5 percent ABV ale. As with Labatt and Molson, there are "Light," "Dry," and "Ice" variants of the Moosehead brand.

Having evaluated the Big Three plus Moosehead, we turn to a brief look at the craft brewing scene which took root in Canada in the 1980s. Horseshoe Bay Brewing, founded by John Mitchell in 1982, was the first microbrewery in Canada and the first brewpub in North America. In Montréal, which eventually evolved into the eastern capital of Canadian craft brewing, the first brewpub, Le Cheval Blanc, opened in 1986 and is still in operation. It has been expanded to include an art gallery and a music venue.

Two of Canada's most important, and still independent, first-generation craft breweries are Big Rock in Edmonton and Calgary, Alberta, founded by Ed McNally in 1985; and Brick Brewing of Waterloo, Ontario, founded by Jim Brickman in 1984. The Big Rock portfolio includes Big Rock "Trad" Traditional Ale, Big Rock Grasshopper Wheat Ale, Big Rock Honey Brown Lager, and Big Rock IPA, all at 5 percent ABV, as well as Big Rock McNally's Extra Strong Ale at 7 percent. Brick products include Brick Red Cap Ale,

Above: Selections from the Wellington Brewery in Guelph, Ontario, Canada's largest independent microbrewery. (Wellington)

## Canadian Beer Festivals

Canada's largest beer festival, the **Kitchener-Waterloo Oktoberfest**, is also the second largest Oktoberfest celebration in the world after the one in Munich, and is attended by a million people. It is held annually in the twin cities of Kitchener and Waterloo in Ontario, beginning on the Friday before Canadian Thanksgiving and runs until the following Saturday.

Founded in 1994, the **Mondial de la Bière** is Montréal's largest beer festival, with an around 80,000 people attending. An international event, it is held in late May and/or early June. **Festibière** was started in nearby Quebec City in 2010. The **Toronto Festival of Beer** (TFOB), founded in 1994, takes

place in August at Exhibition Place in Toronto. It is Canada's second largest festival devoted specifically to beer, and is attended by roughly 30,000 visitors.

Modeled after the GABF in the United States and Britain's GBBF, the **Great Canadian Beer Festival** was founded in 1993 with help from the Victoria Campaign for Real Ale (CAMRA) organization. It is held annually in September in Victoria, British Columbia. In June 2010, **Vancouver Craft Beer Week** became Canada's first annually "beer week." Since then, Ottawa started holding one in June, with Toronto entering the calendar in September.

**Above: Marek Mikunda, the brewmaster at Toronto's Steam Whistle Brewing.** (Steam Whistle)

**Below: Under the Boréale brand name, the ale portfolio of Brasseurs du Nord in Blainville, Quebec includes Blanche, Blonde, Dorée, Rousse, and 6.1 percent ABV Boréale 25.** (Brasseurs du Nord)

Laker Premium Lager, and Brick Waterloo Classic Pilsner, all at 5 percent ABV, and Brick Waterloo Traditional IPA at 5.2 percent.

Meanwhile, Upper Canada Brewing of Toronto was founded by Frank Heaps in 1985 and acquired by Sleeman Breweries in 1998. Upper Canada brews a Dark Ale and Lager, both at 5 percent ABV. Sleeman's product line includes Sleeman Cream Ale and Sleeman Silver Creek Lager, both at 5 percent ABV, as well as 5.5 percent Sleeman IPA and 5.5 percent Sleeman Fine Porter.

Brasseurs du Nord of Blainville, Quebec, founded in 1987, produces the Boréale brands, named for the Aurora Borealis. These now include Boréale Blanche, a 4.2 percent ABV witbier;

6.3 percent Boréale IPA; and 5 percent Boréale Rousse red ale.

The Wellington Brewery of Guelph, Ontario, now Canada's oldest independently owned microbrewery, was founded in 1985 by Phil Gosling as part of the first wave of Ontario microbreweries that began after changes to Ontario law allowed small brewers to operate in the province. Wellington beers include 4 percent ABV Arkell Best Bitter, 5 percent County Dark Ale, 6.5 percent Iron Duke, 4.5 percent Special Pale Ale, 5.5 percent Wellington IPA, and 8 percent Imperial Russian Stout.

Among the other breweries of which the beer traveler may wish to take note while in Canada are Les Brasseurs R.J. in Montréal, a group formed in 1998 through

the merger of three "microbrasseurs," Brasseurs GMT, Brasserie Le Cheval Blanc, and Brasseurs de l'Anse. Located on Mount-Royal Avenue in the heart of the Plateau-Mont-Royal district, R.J. has come to be called the "Official Brewer of the Avenue." Their beers include Belle Gueule Originale, Belle Gueule Pilsner, and Belle Gueule Rousse, all at 5.2 percent ABV, as well as Belle Gueule Hefeweizen and Tremblay, both at 5 percent. Cheval Blanc beers include nearly every imaginable beer style from saison to rauchbier. Some of their beers are 5.5 percent ABV Raizin Weizen fruit beer and 8.8 percent Triple De Mars bière de garde. Saisons include 6.1 percent Kissmeyer Saison Noël and 6 percent Saison Framboise. In addition to their own beers, R.J. brews Belgian d'Achouffe beers under license.

Across the continent in British Columbia, Central City Brewing Company in Surrey is a several times winner of the Canadian

Brewing Awards Brewery of the Year, who earned gold in the World Beer Cup for their 5.5 percent ABV Red Racer ESB, and silver for their 10.25 percent Thor's Hammer Barley Wine. The 6.5 percent Red Racer IPA rates in *Beer Advocate*'s Canadian top ten. Bourbon-Barrel Aged Thor's

**Above: Central City Brewing produces several variations on their award-winning Thor's Hammer barleywine, including this one aged in bourbon casks. (Central City)**

**Below: Toronto's Steam Whistle Brewing is located in a former railroad roundhouse, and is famous for its vintage vehicle collection. (Steam Whistle)**

Above: Inside the Cheval Blanc brew-pub in Montréal. (Cheval Blanc)

Below: The Brasseurs R.J. portfolio includes Belle Gueule Originale, Pilsner, Rousse, and Hefeweizen, as well as Tremblay Blonde. (Brasseurs R.J.)

Hammer Barley Wine tops their ABV list at 12 percent.

In Nelson, British Columbia, the Nelson Brewing Company was originally founded in 1897, folded in the mid-twentieth century, and was revived in 1992 with brew-ing in the original 100-year-old brewery building. In 2006, Nelson decided to "let the brewery evolve to an even more natural state, and by October of that year we were granted Certified Organic status." Nelson products include 4 percent ABV Harvest Moon Organic Hemp Ale, 5 percent After Dark Brown Ale, 4.5 percent Wild Honey Authentic Ale, 6.5 percent Paddywhack IPA, 6 percent Faceplant Winter Ale, 5.7 percent Black Heart Oatmeal Stout, 8.2 percent Full Nelson Organic Imperial IPA, and 5.2 percent Hop Good Session Ale.

Steam Whistle Brewing in Toronto was started by Greg Taylor, Cameron Heaps, and Gregory Cromwell, who met while working together at Upper

Canada Brewing. After that company was sold to Sleeman's and the Toronto brewery was closed, the three friends incorporated as "The Three Fired Guys" and opened Steam Whistle Brewing in March 2000. Heaps and Taylor still run the brewery. The name is derived from steam-powered whistles, which were "icons of the 1950s, a golden era of progress and prosperity, when things were built to last and people enjoyed the simple things in life." Their flagship brand is 5 percent ABV Steam Whistle Pilsner.

On the international scene, Unibroue is one of the most visible of a twenty-first-century generation of innovative, award-winning Canadian brewers. The company was founded in 1990 by André Dion and Serge Racine, who acquired La Brasserie Massawippi the following year. Their "new, cutting-edge microbrewery" opened in Chambly, Quebec, in 1993, and Belgian-trained brewmaster Paul Arnott joined the company in 1999. A distribution network throughout Quebec and across Canada was developed, and exports to the United States and Europe followed. In 2004, the Sleeman Brewery acquired Unibroue, which was in turn acquired in 2006 by Japan's Sapporo.

Three Unibroue beers in the *Beer Advocate* Canadian top ten are 9 percent ABV La Fin Du Monde,

a Belgian-style tripel (10 percent when aged in French oak); 9 percent Trois Pistoles; and 10 percent Unibroue 17 Grande Réserve. Other notable beers are 5 percent Blanche De Chambly, a witbier; 9 percent Don De Dieu; 8 percent Maudite; and the 5.5 percent Éphémère family of fruit beers.

**This page: Unibroue of Chambly, Quebec earns high marks for its beer, as well as for its provocative label art by Asaf Mirza. Fin du Monde (End of the World), a 9 percent ABV triple golden ale, finds the sun shining on Quebec. Maudite (Damned), an 8 percent amber-red ale, has a canoe of early French-Canadian voyageurs traveling through misty skies. (Unibroue)**

# The United States

The story of brewing and beer culture in most nations around the world is deeply rooted in, and affected by, an old and established tradition that continues to exert its influence today. In the United States, on the other hand, we find a situation where tradition was interrupted, suspended for a dozen years, effectively discarded for half a century, then completely reinvented. Since the final two decades of the twen-

tieth century, American brewing and beer culture has been driven by brewers who approached a clean slate with single-minded determination, motivated by their own creativity, a near obsession with innovation, and a consumer base eagerly awaiting the myriad of fascinating new beer brands that are introduced each year.

By the second decade of the twenty-first century, there were more than 2,500 hundred breweries in the United States, a 50-fold increase from 1980. One of this author's greatest regrets in doing this book is that it is impossible to mention, much less describe in detail, more than a fraction of these. For every one that is described, there are many more who are worthy of mention, and worthy of celebration.

The history of brewing in the United States began with indigenous people, especially in the Southwest, but most of our familiar traditions arrived with the European immigrants. The English brought their top-fer-

menting ale yeasts with them and immediately established breweries in the Colonies. By the time of the American Revolution, ales and porters were well developed as part of daily life, and most landowners were as likely to have a brewhouse on their grounds as a stable. Among America's early statesmen, not only Samuel Adams, but both George Washington and Thomas Jefferson were brewers.

Next, around 1840, the first wave of German immigrants introduced bottom-fermenting yeast to American brewing, and the United States soon became a land of lager drinkers.

In 1873, there were 4,131 breweries spread across nearly every small town and city neighborhood, each with its own special style of beer. The idea of a town or neighborhood brewer was as common as the town baker or neighborhood butcher. The advent of a continental railroad network and the invention of ice-making machines made

possible the rise of megabrewers like Schlitz, Pabst, and Anheuser-Busch, who became large regional brewers. By World War II were in a position to launch truly national brands.

In the twentieth century, things changed as nowhere else on earth except for parts of Canada and Scandinavia. Beer simply vanished from the tables of most Americans, and a great industry sputtered to a halt. The temperance movement led to the Eighteenth Amendment of 1920, which instituted Prohibition, a misguided experiment which

**Above: Enjoying beer, music and camaraderie in a painting by John Gannam. It was part of the 1950s "Home Life in America" series used in advertising by the United States Brewers Foundation in a campaign promoting "America's Beverage of Moderation." (Author's collection)**

Above: The Schaefer Brewery of Brooklyn was one of America's five largest in 1948. (Library of Congress)

Below: Serving Hamm's, brewed in St. Paul, at a typical backyard barbecue, circa the 1950s. Hamm's was once a major regional brand. (Author's collection)

decimated an industry and gave rise to organized crime on a heretofore unseen scale. When the curtain of darkness was finally lifted in 1933, only 756 breweries reopened, and many of these failed during the Great Depression. The rise of national brands, with their multi-site industrial-scale breweries, defined the landscape of an industry. Small local and regional brewers could not compete and many disappeared. By 1965, fewer than 200 breweries remained in the United States, and only 15 years later the number of brewing companies had fallen below 50. Meanwhile, the volume of beer produced by the national brands made the United States the largest brewing nation in the world for over half a century. From 105 million hectoliters in 1959, the industry doubled to over 200 million (170 million barrels) by 1979. The corresponding figures for second-place Germany in those years were 46 and 94 million hectoliters.

The recent chronicle of American brewing is a tale of two industries. There is a twentieth-century story of mega-brewers that rose to dominate an industry and rank among the biggest in the world, and a twenty-first-century story of the mushrooming interest in craft brewing by a consuming public interested in artisanal products and in distinctive, innovative cuisine choices. Recent trends in American brewing include higher ABV and higher hop content, including not only a proliferation of IPAs but double IPAs as well. Many American brewers are now barrel aging many of their beers in oak, and it is not uncommon to see unpressurized cask ale offered at brewpubs and specialty beer bars.

Brewery restaurants, with fuller menus than the average brewpubs, were pioneered by brewer Dan Gordon and chef Dean Biersch, who started their first in Palo Alto, California in 1988. Even as the concept proliferated among other restaurateurs, Gordon Biersch evolved

into a multistate operation. Gordon Biersch was later sold to CraftWorks, which also owns the Rock Bottom brewery restaurant chain. Another, similar organization is the BJ's Restaurant & Brewery chain. Dan Gordon still operates a Gordon Biersch production brewery in San Jose, California. Numerous independent brewery restaurants, cropping up all the time, continue to operate nationwide.

Looking forward into the twenty-first century, the chronicle centers on the tale of a craft brewing segment that is growing by double digits annually, and, more important than that, it is controlling the narrative, not only in leading business journals, but among consumers—from the beer traveler visiting America to the "food-ie" beer connoisseurs at home. Such people enthuse about beers such as Anchor Steam, Dogfish Head 90 Minute IPA, or Stone Brewing's Arrogant Bastard, not the flavor-neutral adjunct lagers stacked in supermarket aisles.

The story now is of brewing companies that were unknown in the 1990s when Anheuser-Busch last brewed more than

100 million barrels annually. Looking at the beers and breweries rated most highly by entities such as *Beer Advocate* and *RateBeer*, one sees names such as The Alchemist, AleSmith, Bells, Cigar City, Hill Farmstead, Founders Brewing, Three Floyds, and Toppling Goliath. These names may—or may not—change and be superseded in the pantheon by other names, but this will only speak to the vibrancy of the American craft brewing industry. One name, that of the Russian River Brewing Company, reminds us of their signature beers, Pliny the Elder and the rare Pliny the Younger, which consistently vie

**Above: Pouring a beer for dad. (Library of Congress)**

**Below: One of Milwaukee's greatest names, Schlitz was one of America's "Big Three" for most of the twentieth century. The "cone top" can was introduced in 1935. (Author's collection)**

*Just* THE KISS OF THE HOPS

You don't have to drink a bitter beer. Schlitz methods of brewing control capture *only* the delicate flavor of the hops, not their harsh bitterness. That's one reason for that famous flavor found only in Schlitz. Taste Schlitz and you'll never want to go back to a bitter beer.

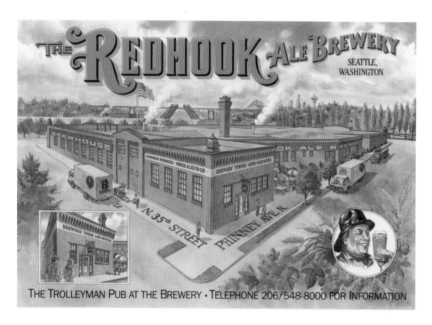

THE TROLLEYMAN PUB AT THE BREWERY · TELEPHONE 206/548-8000 FOR INFORMATION

**Above: The Redhook Ale Brewery, founded in Seattle in 1981 by Paul Shipman and Gordon Bowker, was a pioneer microbrewery that did well.** (Author's collection)

**Below: Classic labels from well-known names of first generation microbreweries which are no longer with us. They include New York City's Manhattan Brewing; Montana Beverages of Helena, known for reviving the Kessler brand; Catamount Brewing of White River Junction, Vermont; and Allan Paul's much missed San Francisco Brewing, that city's first brewpub.** (Author's collection)

with Westvleteren 12 on the lips of consumers as they sip and designate the best beers in the world.

Looking back before the turn of the century, Anheuser-Busch, the industry leader for half a century, controlled 53 percent of the market, and Miller Brewing had a 23 percent market share, with both companies brewing at many sites around the United States. Rounding out the Big Three was Adolph Coors of Golden, Colorado, controlling 13 percent of the market

from the largest single brewery in the world. Within a decade of the turn of the century, they still maintain a volume lead, but their overall numbers are slipping, and their flagship brands are down by double-digit percentages. Furthermore, none of the Big Three is still American-owned. Anheuser-Busch was acquired by Belgian-Brazilian InBev, Miller was acquired by London-based South African Breweries, while Coors is the second name in a consortium in which the first name is that of Canada's Molson.

To complicate the layman's understanding of the corporate structures of Miller and Coors, the United States operations of these two entities formed a joined marketing company called MillerCoors in 2007, intended to share a common sales apparatus in the United States market.

Today, the lead brands of the Big Three still dominate the market. The top two in the United States, and the top brands behind China's Snow worldwide, are Bud Light and Budweiser

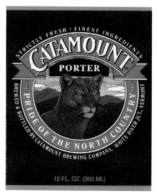

from Anheuser-Busch. In the next tier are Miller Lite and Coors Light as well Anheuser-Busch budget brands such as Natural Light, Busch, Busch Light, and Michelob, the latter once positioned as the Anheuser-Busch premium brand. Miller High Life and Miller Genuine Draft are at the bottom of the top ten, while the onetime flagship Coors (aka Coors Banquet Beer) brand is at the bottom of the top 25. All of them are "adjunct lagers," meaning that they are brewed with rice and corn as well as barley malt, and they range between 4 and 5 percent ABV.

Another company included among the twentieth-century megabrewers is Pabst Brewing. Founded in Milwaukee, America's one-time brewing capital, in 1844 by Jacob Best, and taken over by Captain Frederick Pabst in 1889, Pabst was one of the largest three or four brewing companies in the country for about a century with its flagship lager, Pabst Blue Ribbon. In 1985, Pabst was acquired in a hostile takeover by the enigmatic recluse Paul Kalmanowitz, who began closing breweries and selling assets. As Pabst declined, it became a holding company of bygone brands, buying the fading legacy names of defunct companies such as Heileman, Lone Star, Pearl, Olympia Brewing, Primo, Rainier, Schaefer, Stroh, and Schlitz, which

for many years was, itself, one of the biggest three or four. Since 1996, Pabst has not brewed any beer, but has contracted with other companies to produce its products. In 2010, it was acquired by investor C. Dean Metropoulos, on whose watch the 4.74 percent ABV Pabst Blue Ribbon brand developed a cult following that turned the company's image around. Despite its being a virtual brewer with no breweries, Pabst has more volume in the United States than any brewing company other than the Big Three.

In the 1950-1975 era, as the biggest breweries grew bigger, and the regional breweries faded, mass-market beer gradually became "flavor neutral," designed to appeal to a broader audience,

**Above: Fritz Maytag saved San Francisco's Anchor Brewing from extinction, and became the indispensable role model for the craft brewing movement. (Anchor)**

**Below: The flagship brand for Anchor Brewing is their Steam Beer, which is still their most popular offering. (Anchor)**

Above: Rob and Kurt Widmer started their brewery in 1984, pioneered hefeweizen in American brewing and were part of making Portland, Oregon into a major craft brewing center. (Widmer)

Below: The Deschutes Brewery started by Gary Fish in 1988 as a small brewpub in Bend, Oregon, but it grew into one of America's largest brewers. (Deschutes)

and to do so on the basis of price, not flavor. This redefined American beer culture for at least two generations.

Then, things started to change. Eventually, the craft brewing revolution would redefine the market the other way.

It was in 1965 that a recent Stanford University graduate named Fritz Maytag was given some startling news. The San Francisco brewery that brewed his favorite beer was about to go out of business. Dating back to 1896, the Anchor Brewing Company was typical of the hundreds of local breweries that had survived Prohibition only to be buried by the tide of mass merchandising that had created the national brands. For the price of a used car, Maytag bought a control-

ling interest in Anchor in 1965. The combination of chemistry, business, and tradition had an intoxicating effect on him. "I was made for this brewery," he once told an interviewer. "It was a marriage made in heaven." Though it would take a decade before the business was profitable, Fritz Maytag had seen to it that Anchor did not simply fade away. He also bought himself a career that would define the future of small-scale brewing in the United States.

In 1975 Maytag revived a tradition, long dormant in American brewing culture, of brewing an annual Christmas Beer, priding himself on using a completely new recipe every year. Though he departed from Anchor Brewing in 2010, he left a legacy of serious attention to detail and quality that remains with the company. The products include 4.9 percent ABV Anchor Steam Beer, which has been the brewery's flagship product since before Maytag took over; 5.6 percent Anchor Porter; and 5.9 percent Liberty Ale, which was Anchor's 1976 Christmas Beer.

Inspired by Fritz Maytag, the craft brewery revolution started with the now extinct New Albion Brewing of Sonoma, California, which was founded by Jack McAuliffe in 1976. At the same time that McAuliffe was getting started, there was a growing public awareness of beer styles—such as ales, wheat beers, and stouts—

that were rare or even unknown in the United States.

In 1979, Sierra Nevada Brewing was founded in Chico, California, by homebrewers Ken Grossman and Paul Camusi. It is notable that Sierra Nevada would go on to experience an extraordinary 1,500 percent growth surge during the 1990s that would place it solidly among the top ten American brewing companies.

The craft beer revolution spread to Colorado in 1980 with the opening of the Boulder Brewing Company (Rockies Brewing from 1993 to 2004) and to New York state as William Newman began brewing in Albany the following year. Paul Shipman's Redhook Ale Brewery opened in Seattle in 1981. In Portland, Oregon, Kurt and Rob Widmer, who founded Widmer Brothers Brewing Company in 1984, went on to pioneer an American variation on the German hefeweizen (wheat beer) in 1986. They were also among the first breweries to introduce seasonal beers other than Christmas beers into America.

In the early days, the new craft breweries were called microbreweries, the original definition of which was a brewery with a capacity under 3,000 barrels. By the end of the 1980s this threshold increased to 15,000 barrels as the demand for microbrewed beer first doubled and then tripled.

Another new entity born of the craft brewing revolution was the "brewpub," a microbrewery that both brewed and served its beer to the public on the same premises. Brewpubs had existed in the United States in the eighteenth and nineteenth centuries, but after Prohibition it was illegal in most states to both brew beer and sell it directly to the public on the same site. Changes in local laws since the early 1980s rescinded these outdated restrictions and made it possible for brewpubs to become widespread.

In Yakima, Bert Grant would open America's first brewpub in more than a century in 1982, and others soon followed. Mendocino Brewing, which was established in the appropriately-named village of Hopland, about two hours

**This page: In 1980, Ken Grossman helped start Sierra Nevada Brewing in Chico, California, and grew it into one of America's largest brewing companies. (Sierra Nevada)**

This page: In 1984, Jim Koch founded the Boston Beer Company and created the Samuel Adams brand. Over time, his company became the largest craft brewer in America.
(Boston Beer Company)

north of San Francisco, opened in 1983 and was followed by Buffalo Bill Owens' brewpub in Hayward, across the Bay from San Francisco. By the time that Allan Paul opened San Francisco Brewing in the old Albatross Saloon in 1986, there was a rush of new brewpubs opening throughout the United States. Mike and Brian McMenamin led the brewpub revolution into the Northwest in 1985 with the Hillsdale in Portland, and within four years they had six brewpubs in western Oregon. Today, McMenamins operates 65 brewpubs, microbreweries, music venues, hotels, and theater pubs across Oregon and Washington.

Gary Fish, who established the Deschutes Brewery as a small brewpub in Bend, Oregon, in 1988 is a case in point of the staying power of craft breweries. Said

Fish of the maturing craft brew market of the 1990s, "This industry has more credibility now. . . . It has grown, and the quality of the beer has improved. Failures are just business, not a fad gone awry. We think there are very good times to come for this segment of the industry."

Today, Deschutes is America's twelfth-largest brewer in a much larger field. The American brewing industry that was down to fewer than 50 companies in 1980, had topped 1,500 by 2000 and surpassed 2,500 breweries a decade later.

Despite the disproportional differences in size and market share, the megabrewers took notice of the craft brewers and were befuddled by their popularity, but they did not ignore them. In the 1990s, they even tried to imitate them. Anheuser-Busch introduced an ersatz craft beer called Elk Mountain Amber Ale, and Miller produced its Red Dog. Despite much hype and fanfare, neither survived. Coors was more successful. They even started a "microbrewery," called Sandlot at Denver's major league ballpark. It was here in 1995 that brewer Keith Villa created 5.4 percent ABV Blue Moon Belgian White, a Belgian-style witbier. Coors eventually took the beer national, and it still remains. A decade later, in 2006, Anheuser-Busch rolled out its own 5.2 percent witbier called Shock Top. This brand later became the umbrella for a family of AB InBev seasonal beers which replaced the variants of the venerable Michelob brand in the company's portfolio.

Anheuser-Busch also moved to acquire a minority stake in several craft brewing companies. In 2011, they bought a controlling share of the Goose Island Brewery, started in 1988 in Chicago. Goose Island had become famous in the early twenty-first century for its critically acclaimed 13 percent ABV Bourbon Country Stout.

Among the first craft breweries in the East outside the Empire State were D.L. Geary Brewing in Portland, Maine, and Catamount Brewing in White River Junction, Vermont, which opened in 1986 and 1987 respectively. Another early Eastern micro was Mass Bay Brewing—famous for the Harpoon brand.

On the East Coast, many companies of craft brewing's first generation did not immediately open breweries but contracted with existing regional breweries to produce their beer. Among these was entrepreneur Jim Koch, who formed his Boston Beer Company in 1985 to produce Samuel Adams Lager. His idea was to create an American beer, which he named for the eighteenth-century Boston patriot and maltster Samuel Adams. In 1988, Koch began

brewing at the renovated former Haffenreffer Brewery in an old brewing neighborhood in Boston. By the early 1990s, Sam Adams was available throughout the United States.

With brewing operations in Boston and at a larger facility in Breinigsville, near Allentown, Pennsylvania, the Boston Beer Company produces a sizable number of products, including more than a dozen core beers as well as many annually rotating seasonal beers and many that are brewed on a temporary basis. The flagship products are 4.9 percent ABV Boston Lager (the first Samuel Adams beer, launched in 1984) and 5.4 percent Boston Ale, first released in 1987 to celebrate the opening of the Boston brewery. Other core products include 4.9 percent Cream Stout (since 1993) and 5.5 percent Revolutionary Rye (since 2010). Seasonals include 5.6 percent Winter Lager (since 1989) and 5.3 percent Summer Ale (since

**Above: In 1988, Ed Stebbins and Richard Pfeffer started Gritty McDuff's as a brewpub in Portland Maine. Since then, Gritty McDuff's has opened additional locations in Freeport and Auburn, and has become a Maine institution. (Gritty McDuff's)**

**Below: John Kimmich started the Alchemist Brewery in Waterbury, Vermont in 2003. His Heady Topper has a cult following throughout the Eastern United States and is rated among America's top ten by leading beer enthusiast journals. (Alchemist)**

This page: Scenes from the Allagash Brewing Company of Portland, Maine. Allagash White is the flagship brand. Curieux, an 11 percent ABV tripel, was the "first foray into barrel aging. . . [placed into] bourbon barrels for eight weeks in our cold cellars." (Allagash)

1996). Other beers have included 10.3 percent Infinium, a 2010 limited release of a collaboration with Germany's Weihenstephan. In 2013, in cooperation with Jack McAuliffe, Boston's Jim Koch reintroduced the New Albion brand, the original microbrew, on a limited basis. A 6 percent ABV pale ale, it is said to approximate the 1976 original.

The new microbreweries were not the only institutions that would benefit from Fritz Maytag's bold experiment. Among the other beneficiaries of the interest in craft brewed beer were the other independent regional brewers who, like Anchor, had been bucking the consolidation trend for decades. Just as beer drinkers were discovering a new generation of brewers who were recapturing beer and brewing styles that were an homage to the brewers of their grandfathers' day, others were enjoying beer from the same breweries that had supplied their grandparents. A case in point is the brewery founded in 1829 by David Gottlob Jüngling in Pottsville, Pennsylvania. D.G. Yuengling Brewing is now the oldest brewery in the United States. In 1985, as the craft brewing movement was just beginning to sweep the country, Richard L. "Dick" Yuengling bought the family "microbrewery" from his father. The brewery's success was a corollary to the success of the craft brewing movement.

Still headquartered in Pottsville, Yuengling brews there as well as in Tampa, Florida. Widely distributed in the Eastern United States, Yuengling's products include 4.4 percent ABV Yuengling Traditional Lager, 5.4 percent Lord Chesterfield Ale, and 4.7 percent Yuengling Porter, which was in the portfolio for many decades in the twentieth century before craft brewers rediscovered porter as a style.

Genesee Brewing (known as High Falls Brewing between 2000 and 2009) of Rochester, New York, was one of numerous fading regional breweries in the 1980s when it reinvented itself as a contract brewery of other companies' brands. In 2009, it was acquired by the investment firm KPS Capital Partners, who rolled it into a holding company called North American Breweries, to which they added the United States operations of Canada's Labatt. In 2012, this entity was sold to Costa Rica–based Florida Ice & Farm, the parent company of Cervecería Costa Rica.

Also part of the holdings of North American Breweries at the time was Magic Hat Brewing, a craft brewery started in 1994 in Burlington, Vermont. In 2008, Magic Hat had acquired Pyramid Breweries of Seattle. Originally started as Hart Brewing in 1984, Pyramid had grown into a large regional brewing company, with brewery restaurants across the Pacific Northwest and California, and in 2004 had acquired the Portland Brewing Company, makers of the MacTarnahan's brand.

In the twenty-first century, with Pabst as a virtual brewer and the Big Three plus Genesee now foreign owned, Yuengling, Koch's Boston Beer, and Grossman's Sierra Nevada became the three largest brewing companies owned in the United States who actually brew beer in the United States. The annual output for both Yuengling and Boston Beer is roughly tied at around 2.5 million barrels, while Sierra Nevada produces about 800,000 barrels.

Sierra Nevada has brewed at the same plant in Chico since 1989 and added a new facility in Mills River, North Carolina, in 2014, which will greatly expand its market presence in the East. The flagship product since the beginning has been 5.6 percent ABV Sierra Nevada Pale Ale. Other products dating from the early days include 5.6 percent Sierra Nevada Porter and 5.8 percent Sierra Nevada Stout. Another product dating to the 1980s, and now a seasonal "special release," is 9.6 percent Bigfoot Barleywine Style Ale. Two products released in 2009 which earned high marks from critics were 7.2 percent Torpedo, an IPA, and 6.7 percent Chico Estate Harvest Ale, originally released in 2008, which was

made with organic hops and barley grown on land that is part of the brewery premises.

Across the continent, meanwhile, one finds an especially high concentration of highly regarded craft brewers in New England. John and Jen Kimmich opened The Alchemist in Waterbury, Vermont, in 2003 as a brewpub and began brewing a range of beers. One of the beers that attracted special attention was their unfiltered 8 percent ABV Heady Topper double IPA, which became so popular that they began packaging it for retail sale. Thought it is unavailable outside the immediate area, it has attracted worldwide attention. In 2013, *Beer Advocate* rated it as the number-one beer in the world, calling it "a complex web of genius."

Allagash Brewing, founded in Portland, Maine, in 1994 by Rob Tod, is best known for their highly regarded flagship brand, 5 percent ABV Allagash White, a widely distributed "interpretation of a traditional Belgian wheat beer. Brewed with a generous portion of wheat and spiced with

This page: Shaun Hill's Hill Farmstead Brewery on the family's Vermont farm was rated by the *RateBeer* website as the best brewery in the world in 2013. (Bob M. Montgomery photos)

**Above: The first of several Heartland Brewery locations opened on Union Square in New York City in 1995.** (Bill Yenne photo)

**Below: Filling barrels for aging at the Rock Art Brewery in Morrisville, Vermont.** (Rock Art)

coriander and Curacao orange peel." Other beers include their 7 percent Dubbel Ale, their 9 percent Tripel Ale, and 11 percent Allagash Curieux, made by aging Tripel Ale in bourbon barrels for eight weeks in cold cellars. The aged beer is then blended back with a portion of fresh Tripel.

Gritty McDuff's, also in Portland, started in 1988 as a brewpub in which this author spent a memorable snowy evening about a decade later. Another pub with a larger brewery opened in Freeport in 1995, and a third brewpub was added in Auburn in 2005. The bottled product portfolio includes Original Pub Style, which is based on the early draft beers. Other products include 5 percent ABV Maine's Best IPA and 5 percent Gritty's Best Bitter as well as seasonal 6 percent Halloween Ale, 6.2 percent Christmas Ale, and 4.9 percent Vacationland Summer Ale.

The Hill Farmstead Brewery, located at the end of a dirt road near Greensboro, Vermont, is a true farmhouse brewery that happened to be named as the 2013 Best Brewery in the World by *RateBeer*, who listed eight of their offerings among its top ten new beers in the world. Owners Shaun and Darren Hill name the beers of their popular "Ancestral Series" after their own ancestors, who have lived on the same farm for more than two cen-

turies. Ephraim, for example is named for great-great grandfather Ephraim Hill, who was born there in 1823. The brewery is several hundred feet downhill from the land that he and his father settled. The beer is a 9.8 percent ABV unfiltered and double dry hopped imperial IPA. Topping *RateBeer*'s list is Ann, a 6.5 percent saison aged in French oak wine barrels, and Society & Solitude #2, a 9.5 percent black ale.

Matt Nadeau started his Rock Art Brewery in 1997 in his Johnson, Vermont, basement, but moved to a larger brewery in nearby Morrisville in 2001. The beers include 7.2 percent ABV Rigderunner, 5 percent Whitetail, 8 percent Double IPA, and a 10 percent barleywine called Vermonster.

In upstate New York, many breweries have come and gone since William Newman's pioneering efforts in the 1980s. Beginning in the 1990s, brewpubs such as Big House Brewing and Malt River Brewing came and went, while Brown's Brewing and the Albany Pump Station came and stayed. Scott Vaccaro's Captain Lawrence Brewing in Elmsford, New York, came to stay as well. Named for Captain Samuel Lawrence, who served in the Westchester County Militia in the American Revolution, the brewery's lineup includes 9 percent Captain's Reserve Imperial IPA.

New York City's first, and long departed, craft brewing companies included Manhattan Brewing, a brewpub started by Robert D'Addona in 1984 in a former Con Edison substation in Soho, as well as Old New York Brewing, known for its Amsterdam Amber, which was brewed under contract by F.X. Matt Brewing in Utica, in upstate New York.

F.X. Matt, also known as West End Brewing, would evolve into one of the largest of the contract brewers in the East during the late 1980s and early 1990s. Originally founded in 1853, the company that had long been famous for its own Utica Club brand would contract brew for a number of eastern companies during the 1980s and 1990s.

New York City has seen the comings and goings of a number of brewpubs since Manhattan Brewing, including the now defunct Zip City and the Heartland Brewery chain, which started in 1995 and has grown to multiple Manhattan sites. Heartland brews for these locations at a production facility in Brooklyn. Best known of the craft brewers in that borough is the Brooklyn Brewery, started in 1987 by Steve Hindy and Tom Potter. Their beer was brewed upstate at F.X. Matt until they acquired a brewery site in Brooklyn in 1996. As a brewmaster, they

hired Garrett Oliver, formerly with Manhattan Brewing and the author of *The Brewmaster's Table: Discovering the Pleasures of Real Beer with Real Food*. The Brooklyn beers include 4.5 percent ABV Brooklyn American Ale, 10 percent Brooklyn Black Chocolate Stout, and 4.5 percent Blanche De Brooklyn.

Down the shoreline, the Tun Tavern in Atlantic City is an example of a single-site brewery and restaurant with an enviable location. Named for Philadelphia's historic eighteenth-century Tun Tavern, it opened in 1998.

In Pennsylvania, the first generation of craft brewers, circa 1986, included Jeff Ware and Rosemarie Certo of Dock Street in Philadelphia and Tom Pastorius, who started Pennsylvania Brewing, aka Penn Brewery, in Pittsburgh. Both contracted production elsewhere until 1989, when brewpubs were legalized in the state, and then both opened as their own facilities. Dock Street existed as both a brewpub and an offsite bottling operation until 2000.

Jeff Ware moved overseas, but Rosemarie Certo reacquired the brand and reopened Dock Street as a brewpub in 2007. Pastorius restored the former Eberhardt & Ober Brewery. The company was acquired by an

Above: **Rosemarie Certo of Dock Street Brewing.** (Dock Street)

Below: **Joe Coffee Porter and Walt Wit witbier from Philadelphia Brewing.** (Philadelphia Brewing)

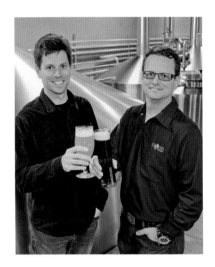

Above: Chris and John Trogner of Tröegs Brewing in Pennsylvania. (Tröegs)

Below: The Dogfish Head portfolio includes 90 Minute IPA, Indian Brown Ale, 60 Minute IPA, Midas Touch, and Palo Santo Marron. All are high in both ABV and IBU (International Bitterness Units). (Dogfish Head)

investment firm in 2003 and sold to other investors in 2009.

Part of a newer generation is Philadelphia Brewing, founded in 2007 by Bill Barton, Nancy Barton, and Jim McBride. Brewing was originally outsourced, but operations are now in the nineteenth-century Weisbrod & Hess Oriental Brewing building, where they brew 5 percent ABV Joe Coffee Porter and 4.2 percent Walt Wit witbier.

Tröegs Brewing was founded in 1996 in Harrisburg, Pennsylvania, by John and Chris Trogner, but relocated to larger facilities in Hershey in 2011. The product line includes a larger than typical proportion of seasonal beers. Year-round beers include 4.8 percent ABV

Dreamweaver wheat ale and 8.2 percent Troegenator double bock. Autumn-December beers include 11 percent Mad Elf, and spring beers include 9.3 percent Flying Mouflan barleywine.

The Dogfish Head Brewery was started in 1995 in Milton, Delaware, by Sam Calagione, but it was in the twenty-first century that the popularity of the beer and the notoriety of the brand became the subject of national media attention. Between 2003 and 2006, the brewery experienced a 400 percent growth spurt, and it has been featured in several television documentaries including the Discovery Channel's *Brew Masters* series. Calagione started with a brewpub in Rehoboth Beach and has licensed the brand

to Dogfish Head Alehouse brew-pubs in Maryland and Virginia. He is also the author of several books on the business of brewing.

Calagione is part of the "extreme beer" school of brewing, which experiments with large quantities of non-traditional ingredients, high ABV, and extreme quantities of bittering hops. Dogfish Head's flagship product line is a series of IPAs whose names indicate the boil time during which hops are continuously added to the wort while brewing. This group includes the 6 percent ABV 60 Minute IPA (rated at 60 IBU), introduced in 2003, and 9 percent 90 Minute IPA (90 IBU), introduced in 2001. There is also a limited-release 18 percent 120 Minute IPA with a bitterness of 120 IBU. Other year-round brands are 9 percent Midas Touch, introduced in 1999 and made with honey, white muscat grapes, and saffron; 12 percent Palo Santo Marron, which is an unfiltered brown ale fermented in Paraguayan palo santo wood; and 8 percent Raison D'Etre, a Belgian strong dark ale brewed with beet sugar and raisins. Dogfish Head has also brewed beer in collaboration with other brewers, such as Epic Brewing of New Zealand, in which the beer was fermented with tamarillo (a tomato-like fruit also called the tree tomato) which had been smoked using wood chips from the Pohutakawa tree.

Above: Sam Calagione of Dogfish Head. (Dogfish Head)

Below: Tap handles at the Flying Dog Brewery. (Flying Dog)

Other collaboration partners have included Stone Brewing, Three Floyds, and Victory Brewing. For the 5.5 percent Repoterroir in 2011, Dogfish Head teamed with Sierra Nevada, Avery Brewing, Allagash Brewing, and Lost Abbey.

The Flying Dog Brewery started as a brewpub in Aspen, Colorado, in 1990, acquired a production facility in Denver in 1994, and purchased Frederick Brewing in Frederick, Maryland, in 2006. In 2008, the Denver plant closed and production was moved east. Of note were the labels, designed by illustrator Ralph Steadman. Products include 5.5 percent ABV Doggie Style Classic Pale Ale, 8.3 percent Raging Bitch Belgian-Style IPA, and 9.2 percent Gonzo Imperial Porter.

Above, both: Scenes from the award-winning Blue Mountain Brewery in Afton, Virginia. Started in 2007, it is part of the "Brew Ridge Trail," a series of craft breweries along the along the Blue Ridge Parkway.
(Blue Mountain)

Below: Beers from Cigar City Brewing of Tampa, Florida include 9 percent ABV Bolita Double Nut Brown Ale, 11 percent Cognac Barrel Aged Imperial Sweet Stout, and 7.5 percent Humidor Series IPA.
(Cigar City)

In Virginia, the state's original large-scale craft brewer, Old Dominion, was relocated to Delaware in 2006 after 17 years, but around that time a series of new brewery restaurants were established along the Blue Ridge Parkway. The first of these was the Starr Hill Brewery, which moved from Charlottesville to Crozet in 2005. In 2007, the Blue Mountain Brewery opened in Afton, and in 2008 Devils Backbone Brewing Company arrived in Roseland. Blue

Mountain products include 5.9 percent ABV Full Nelson Virginia Pale Ale and 10 percent Dark Hollow Artisanal Ale.

Cigar City Brewing in Tampa was, according to its mission statement, "founded with two goals in mind. The first to make the world's best beer and the second to share with people near and far the fascinating culture and heritage of the Cigar City of Tampa . . . [f]rom its past as the world's largest cigar producer to its Latin roots." Cigar City con-

sistently wins awards and critical acclaim. In 2012, it was named as the number four "Best Brewer in the World" by *RateBeer*. Among its most highly rated beers are 11 percent ABV Hunahpu's Imperial Stout, 11 percent Marshal Zhukov's Imperial Stout, 8.5 percent Good Gourd Imperial Pumpkin Ale, 7.5 percent Jai Alai IPA, 9 percent Bolita Double Nut Brown Ale (named in honor of the popular Bolita games), 11 percent Cognac Barrel Aged Imperial Sweet Stout, and 7.5 percent Humidor Series IPA.

Back in the 1970s, in the heyday of Willie Nelson and the "Outlaw" movement in country music, San Antonio had been the brewing capital of the Lone Star State and the "National Beer of Texas" was San Antonio–brewed Lone Star. However, the brewery soon went through a revolving door of ownership before being closed permanently in 1996. The leading beer in Texas today is a brand that has been around since 1909 and is brewed by one of the top ten brewing companies in the United States. This is Shiner, from the Spoetzl Brewery in Shiner, Texas, south of San Antonio. Founded by Kosmos Spoetzl, a German who once brewed in Egypt, the brewery was not on the national radar until Carlos Alvarez of San Antonio bought the company in 1989 and increased sales eight-fold by the early twenty-first

century. Today, the "National Beer of Texas" is probably 4.4 percent ABV Shiner Bock, a beer that his been around for decades and is finally getting its due. Other beers include 5.5 percent Shiner Hefeweizen and 4.9 percent Shiner Kosmos Reserve.

Boulevard Brewing, who began marketing its beer in 1989, grew into the largest specialty brewer in the Midwest by the twenty-first century. Founded by John McDonald, Boulevard underwent small expansions in 1999 and 2003 and a major enlargement of operations in 2005 that gave it a 700,000-barrel annual capacity for its 5.4 percent ABV Pale Ale, 4.4 percent Unfiltered Wheat Beer, 6 percent Bully! Porter, 8.5 percent Tank 7 Farmhouse Ale, 8.5 percent Double-Wide IPA, and various seasonal beers.

**This page: Spoetzl Brewery's Brewmaster Jimmy Mauric on the bottle line, and Shiner Bock, the flagship brand. (Spoetzl)**

northern Midwest. Its flagship is 4.8 percent ABV Original Deer Brand lager, complemented by 5 percent FireBrick. Seasonal beers include 4.4 percent Hefeweizen and 5.1 percent Schmaltz's Alt. The Grain Belt portfolio includes 4.6 percent Grain Belt Premium and 4.7 percent Grain Belt Nordeast. Schell recently released The Star of the North, their 3.5 percent interpretation of a traditional Berliner Weisse.

Another Minnesota brewery of note is Summit Brewing, started in St. Paul in 1986 and now the state's second largest. They are known for their 5.5 percent ABV Extra Pale Ale, 6.4 percent IPA, and 6.6 percent Oktoberfest.

Like Schell, the Stevens Point Brewery, aka "Point," has been brewing since the nineteenth century. Founded in Stevens Point,

**Above: KBS (Kentucky Breakfast Stout) and Double Trouble IPA from Founders Brewing. KBS is rated by beer enthusiast journals in America's top ten. (Founders)**

**Below: Leading Midwest brands include Deer Brand and Schmaltz's Alt from Schell; Great Northern Porter and Pilsner from Summit Brewing, and Unfiltered Wheat from Boulevard in Kansas City. (Courtesy of the breweries)**

August Schell Brewing in New Ulm, Minnesota, was started in 1866 by August Schell, survived the ups and downs of industry consolidation in the twentieth century, and is today the second-oldest family-owned brewery in the United States after Yuengling. Having acquired the classic Grain Belt brand in 2002, it is now the largest brewing company in the

Wisconsin, in 1857 by George Ruder and Frank Wahle, Point provided beer for troops during the Civil War and went through a number of private owners before being sold to Barton Beers, a Chicago-based distributor, in 1992. Since being resold in 2002 to Milwaukee-based investors, it has acquired outside brands, including Augsburger from the late megabrewer Stroh, and those of the Wisconsin microbrewery James Page Brewing, which existed from 1986 to 2005. Once available only within a narrow radius of Stevens Point, the products are now sold in most states. These include 4.7 percent ABV Classic Amber, 5.38 percent Belgian White, 5.2 percent Black Ale, 5.4 percent Cascade Pale Ale, 6.5 percent Beyond The Pale IPA, 3.5 percent Drop Dead Blonde, 5 percent Burly Brown, 3.65 percent Horizon Wheat, 4.7 percent Amber Classic, and the iconic 4.5 percent Point Special Lager.

Founders Brewing Company in Grand Rapids, Michigan, produces some of the most highly-rated beers in the Midwest. Their

**Above: Part of the portfolio from the Stevens Point Brewery in Wisconsin. Dating back to 1857, it is one of the oldest privately-held breweries in the United States. (Stevens Point)**

**Below: The "World's Largest Six Pack" in LaCrosse, Wisconsin is a group of a half dozen fermenters built in 1969 by G. Heileman, and now owned and operated by the City Brewery. (Bill Yenne photo)**

This page: Kim Jordan of Colorado's New Belgium Brewing, and highlights from her portfolio. (New Belgium)

11.2 percent ABV Founders KBS (Kentucky Breakfast Stout) is rated by *Beer Advocate* among the top ten in the United States. *RateBeer* named Founders as the

third-best brewery in the world in 2013. Founders was founded in 1997 as Canal Street Brewing by homebrewers Mike Stevens and Dave Engbers. They named the brewery after the street where many early breweries had been and pictured these on their labels with the word "Founders" above. This name caught on, and the company name changed. Among their other highly rated beers are 8.3 percent Founders Breakfast Stout, 10.5 percent Founders Imperial Stout, an 9.4 percent Founders Double Trouble IPA.

New Belgium Brewing in Fort Collins, Colorado, well known for its 5.5 percent ABV Fat Tire Amber Ale, is the second-largest brewer in the Mountain West, after Coors, and the eighth larg-

est in the United States. The story began in 1991 when homebrewer Jeff Lebesch began operating commercially. On a trip across Belgium on a fat-tired bicycle, Lebesch was inspired by the imaginative beers that he discovered in Belgium—hence the brewery name and the brand name. In the early days, it was a family business, with Lebesch's wife, Kim Jordan, as part of the team. They have since divorced, Lebesch left the company in 2001, and Jordan is now chief executive. The product line includes 5.6 percent 1554 Enlightened Black Ale, 7 percent Abbey Belgian Style Ale, 6.5 percent Ranger IPA, 6.2 percent Snow Day Winter Ale, and 7.8 percent Trippel Belgian Style Ale.

Also in Fort Collins, but less than 10 percent the size of New Belgium, is Odell Brewing, started in 1989 by Doug Odell, his wife Wynne, and his sister Corkie. Their beers include 4.2 percent ABV Easy Street Wheat, 5 percent Levity Ale, 7 percent IPA, 5.2 percent 5 Barrel Pale Ale, 5.3 percent 90 Shilling Ale, 5.5 percent 90 Oak-Aged Shilling

**Above: Highlights from among the extensive offerings from Odell Brewing. (Odell)**

**Below: The Odell Brewery in Fort Collins, Colorado. (Odell)**

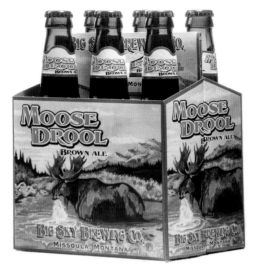

Above: A six-pack of 5.3 percent ABV Moose Drool Brown Ale, the flagship product from Big Sky Brewing of Missoula, Montana, and the man who brews it, head brewer Matt Long. (Loren Moulton, Big Sky)

Below: Products from Great Northern Brewing of Whitefish, Montana include 4.75 percent Wheatfish Wheat Lager, 5.9 percent Snow Ghost Winter Lager, and 5.7 percent Going To The Sun IPA. (Great Northern)

Ale, 5.5 percent Cutthroat Porter, and 9.2 percent ABV Myrcenary Double IPA.

Montana, meanwhile, has had a long history of craft brewing, beginning with Montana Beverages of Helena, which brewed between 1982 and 1999 under the fondly remembered brand of the Kessler Brewing Company, which itself had existed in Helena from 1865 to 1958. They also produced products for craft brewing companies throughout the West. Helena is now home to Blackfoot Brewing and Lewis & Clark Brewing. Another

early Montana brewer is Bayern Brewing, started in Missoula in 1987 by Thorsten Geuera and Jürgen Knöller and now the oldest operating brewery in Montana. The state's largest brewery is Big Sky Brewing started in Missoula in 1995 as a microbrewery by Neal Leathers, Bjorn Nabozney, and Brad Robinson. It is now a major regional brewery with sales in much of the United States. Its flagship brand is 5.3 percent ABV Moose Drool Brown Ale, while other beers include 4.7 percent Scape Goat Pale Ale, 6.2 percent Big Sky IPA, 7.2 percent Powder

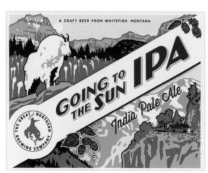

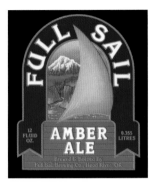
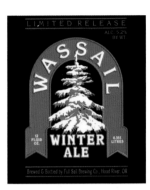

Hound Winter Ale, 4.8 percent Summer Honey, and 4.7 percent Montana Trout Slayer Wheat Ale.

Another notable Montana brewery is Great Northern Brewing in Whitefish, started in 1995 by Minott Wessinger, the great-great grandson of Oregon brewing legend Henry Weinhard. The flagship beer is 4.5 percent ABV Black Star, a double-hopped golden lager. Production of this beer was suspended in 2002 when Wessinger left the company to pursue other interests. Both Wessinger and Black Star returned in 2010. Other products include 4.75 percent Wheatfish Wheat Lager, 4.6 percent Wild Huckleberry Wheat Lager, 5.7 percent Going To The Sun IPA, and 5.9 percent Snow Ghost Winter Lager, a Munich-style dunkel.

The Pacific Northwest quickly became one of the most important—or, by local reckoning, the *most* important—centers of craft brewing in the Western Hemisphere. Back in the twentieth century, the Northwest had been home to three major

household-name regional breweries whose geographic scope was much greater than most strictly regional brewing companies east of the Rockies. Each of them withered late in the century under competition from national brands. Blitz-Weinhard in Portland, Oregon, which originated with Henry Weinhard in 1856, was closed in 1999 after acquisition by Miller Brewing. Olympia Brewing Company dates back to the brewery established in 1896 in Tumwater, Washington, by

**Above: Classic labels of some of the early craft beer brands in the Pacific Northwest include Seattle's Redhook ESB, Full Sail Amber Ale and Full Sail Wassail Winter Ale, and Dead Guy Ale from Rogue Ales of Newport, Oregon. All are still brewed, though labels have changed. (Author's collection)**

**Below: Rogue Brewmaster John Maier with Chuck Linquist and Rogue founder Jack Joyce at Rogue's Bayfront brewpub in Newport in 1991. (Bill Yenne photo)**

Above: Brewmaster Cam O'Connor at the Deschutes Brewery. (Deschutes)

Below: Popular brands, including the flagship. As the company notes "with a dark beer as our first and flagship brand, Black Butte defined Deschutes as a radical player." (Deschutes)

Leopold Schmidt, a German immigrant from Montana. Sold to Pabst in 1983, it later passed to Miller and was closed in 2003. The Rainier brand from Seattle dates back to 1884 at a brewery started in 1878. The Rainier brand and brewery was acquired in 1935 by Emil Sick, who used it as the flagship of an international empire of breweries stretching from Oregon to Montana to Alberta that survived until the 1970s. The Seattle brewery was later sold to Pabst and closed in 1999.

Fast forward to the craft brewery revolution—the next big grouping of Northwest names included Redhook in Seattle (1991); Full Sail in Hood River, Oregon (1987); Deschutes in Bend, Oregon (1988); Rogue in Newport,

Oregon (1988); and a large group of breweries in Portland. These included BridgePort (1984), Portland Brewing (1986), and Widmer Brothers (1994). In 2008, two of the largest of these, Redhook (now with breweries Woodinville, Washington and Portsmouth, New Hampshire) and Widmer, merged to form the NASDAQ-traded Craft Brew Alliance (CBA).

While brewing and brands remained distinct, the joint venture, especially though a minority ownership by AB InBev, gave them a substantially improved nationwide marketing and distribution network. In 2010, Kona Brewing joined the CBA. Today, the CBA ranks ninth among all American brewing companies and Deschutes is in twelfth place.

Redhook brands include 5.8 percent ABV Redhook ESB, 6.2 percent Redhook Long Hammer IPA, and 6 percent Redhook Winterhook.

## American Beer Festivals

**The Great American Beer Festival** (**GABF**), hosted annually in Denver by the Brewers Association, the American craft brewers' trade group, is the largest paid-admission beer festival in the United States. It has an annual attendance of close to 50,000 patrons sampling roughly 3,000 beers from 500 American breweries. Started in 1982 by Charlie Papazian, the three-day event has grown in importance through the years as the gold, silver and bronze medals awarded in more than 80 style categories are considered among the brewing community as the definitive measure of quality. The Brewers Association also hosts the **World Beer Cup**, an international beer competition heavily weighted to American brewers.

One of the biggest and most popular American regional beer festivals is the **Oregon Brewer's Festival**, held in July since 1988 in Portland's Tom McCall Waterfront Park.

It is attended by more than 60,000 people. Other important festivals include the **San**

(Festival photo of J.R. Hubbard and Mike Bolos by the author. Commemorative coasters from the author's collection)

**Francisco International Beer Festival**, started in 1983 and held each May, and New York City's **Brewtopia Great World Beer Festival**, which was started in 2002 and is held in October. Also in the East, the **Mid-Atlantic Beer Festival**,

which began in 2001, has been held both Baltimore and Washington, DC.

Various cities around the United States also play host to **Oktoberfest** celebrations, with some of the largest and best known being those in Cincinnati, Ohio; San Francisco; Denver and Leavenworth, Washington.

In recent years, the event calendars in many American cities have come to include an annual "**Beer Week**," often sponsored by local brewers' organizations. Cincinnati, New York, Grand Rapids and San Francisco hold theirs in February; Chicago, San Antonio, Seattle, Madison, Wisconsin, and Asheville, North Carolina all have their Beer Weeks in May; Philadelphia's is in June; Austin's is around the end of October; and Houston's is in November.

Above: A wide-angle look into the busy inner workings of Seattle's Pike Brewery. (Pike Brewery)

Below: Some highlights from the Pike portfolio include 6.5 percent ABV Pike IPA, 7 percent Pike XXXXX Extra Stout, 9 percent Monk's Uncle Tripel, 9.9 percent Pike Old Bawdy Barley Wine, and the annual 5.5 percent Auld Acquaintance Hoppy Holiday Ale. (Pike Brewery)

Widmer's product line is headed by its flagship 4.9 percent Widmer Hefeweizen, the most visible wheat beer on the West Coast, as well as 5.7 percent Drifter Pale Ale and 6.5 percent Pitch Black IPA.

The Deschutes portfolio is headed by 5.2 percent ABV Black Butte Porter and 5 percent Mirror Pond Pale Ale but also includes

11 percent The Abyss, a highly regarded imperial stout.

Having dwelt upon the Northwest's new generation of mega-brewers, it should be mentioned that the region is literally alive with such a wonderland of microbreweries and brewpubs that a beer traveler could not exhaust them, even on an extended visit. There is the Bayfront Brewery in Newport, Oregon, the showplace of Rogue Ales (and fond memories for this author) since 1989. A traveler might stop here for a 6.5 percent ABV Dead Guy Ale, a 6 percent Smoke Ale, or a 6.1 percent Shakespeare Oatmeal Stout, one of Rogue's original beers.

Also overlooking the salt water is Diamond Knot Brewing, a brewpub started back in

These pages: Inside the pub at the Pike Brewery, with founders Rose Ann and Charles Finkel, and a pasta pairing featuring Pike Pale Ale. (Pike Brewery)

1994 in Mukilteo, Washington. According to the *Seattle Times*, they are "widely regarded as producing some of the best, most innovative beers in the region." In Seattle itself, the traveler may find Georgetown Brewing, a brewpub started in 2002 by Manny Chao and Roger Bialous.

If the beer traveler has time for but one Seattle stop, it should be the Pike Pub & Brewery in the Pike Place Market. It was started in 1989 by Charles and Rose Ann Finkel, who, though their company, Merchant du Vin, were among the first importers of select specialty beer from Europe. As the company's official fact sheet points out, Merchant du Vin "became the exclusive agent for some of Europe's finest independent brewers including Ayinger, Lindemans, Melbourn Bros, Orval, Pinkus, Samuel Smith and

Above: At Seattle's Georgetown Brewing, the tap handles and a brewer adding hops to the brew. (Georgetown)

Below: Serving up a Blonde at Diamond Knot in Mukilteo, Washington. (Diamond Knot)

Traquair House. Their's was the first company to offer a range of Belgian beers; to work with British brewers to create long forgotten styles like Oatmeal Stout, Porter, Imperial Stout and Scotch Ale; and to repackage [their Celebrator, a] classic Bavarian doppelbock under a private label that eventually became the Ayinger Brewery's world-wide brand. They introduced many people including many early craft brewers, to the glories of great beers." Indeed, this is how this author first came to know them.

Charles Finkel is one of the most important men in American beer connoisseurship in the past generation. Among other things, he was also a pioneer in the art of beer and food pairings. He is remembered as the man who convinced the Plaza Hotel in New York City to introduce a beer list to complement their wine list. This was in the 1980s, well before it was as common as it has become in the twenty-first century.

The Finkels, in turn, founded their own brewery in Seattle in 1989. Some of the beers which they and their crew at Pike serve up are 6.5 percent ABV Pike IPA, 5.3 percent Pike Pale, 7 percent Pike XXXXX Extra Stout, 9 percent Monk's Uncle Tripel, 9.9 percent Pike Old Bawdy Barley Wine, and their annual 5.5 percent Auld Acquaintance Hoppy Holiday Ale.

At the extreme corner of the Northwest, Alaskan Brewing in Juneau is still owned and managed by Marcy and Geoff Larson, who founded the brewery in 1986.

This page: These scenes from Alaskan Brewing of Juneau include founders Geoff and Marcy Larson on the bottle line, and product shots of their limited release 6.5 percent ABV (and 45 IBU) Smoked Porter, launched in 1988, and the flagship 5.3 percent Alaskan Amber. (Alaskan Brewing)

Above: Brewmaster Dylan Schatz with a bottle of his Steelhead Extra Pale Ale. (Mad River Brewing)

Below: Long-time portfolio highlights from the Lost Coast Brewery in Eureka, California. (Lost Coast)

The long-time flagship, 5.3 percent ABV Alaskan Amber, an alt beer, heads a list that includes 5.3 percent Freeride American Pale Ale, 6.2 percent Alaskan IPA, and limited releases such as 6.4 percent Alaskan Winter Ale and 6.5 percent Alaskan Smoked Porter.

In California, where craft brewing was born with Fritz Maytag, Jack McAuliffe, and Ken Grossman, there are literally dozens of microbreweries and brewpubs from one end of the state to the other. On a drive south from Oregon, one passes through Humboldt County, home to Mad River Brewing in Blue Lake and Lost Coast in Eureka. Mad River was started in 1989 and today brews a variety of beers including its flagship 5.6 percent ABV Steelhead Extra Pale Ale, 6.5 percent Steelhead Scotch Porter, 6.6 percent Steelhead Extra Stout, 8.6 percent Steelhead Double IPA, and 6.5 percent Jamaica Brand Red Ale.

Lost Coast was opened as a brewpub and café in 1990 by homebrewers Barbara Groom and Wendy Pound in the former meet-

ing hall of the Fraternal Order of the Knights of Pythias. The production brewery was moved offsite in 2005. Influenced by British brewing traditions, the portfolio includes 5 percent ABV Downtown Brown, 4.8 percent Great White, and 5.5 percent Alleycat Amber.

Down the coast in Fort Bragg, one finds North Coast Brewing, which opened in 1988 as a local brewpub. Since that time, brewmaster Mark Ruedrich has transformed North Coast into a major regional brewery with distribution to most of the United States, Europe, and beyond. Among their beers are 9 percent ABV Old Rasputin Imperial Stout, 7.6 percent PranQster Belgian Style Golden Ale, and 9.4 percent Brother Thelonious Belgian Style Abbey Ale. The latter is a reference to jazz pianist and composer Thelonious Monk, and a pun on his surname because abbey beers are associated with monks. North Coast, meanwhile, is a major sponsor of the Monterey Jazz Festival, SFJazz, the Mendocino Music Festival, and the Oregon Jazz Party.

North Coast's 7.9 percent Le Merle, a rustic Belgian-style saison took a gold medal at the Brussels Beer Challenge in 2012.

Were the beer traveler to begin the trek at the opposite end of California, the first stop would naturally be San Diego, which

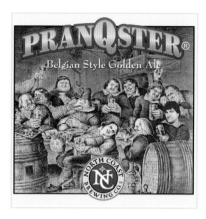

has developed a beer culture to rival those of San Francisco and the Pacific Northwest. One of the earliest fixtures on the San Diego craft brewing scene was Karl Strauss Brewing, founded in 1989 as a brewery and German restaurant by Chris Cramer, Matt

**Above: Brother Thelonious and PranQster, two Belgian-style ales from North Coast Brewing of Fort Bragg, California. (North Coast)**

**Below: Pouring pints at Karl Strauss Brewing in San Diego. (Karl Strauss)**

**Above: Greg Koch and Steve Wagner, founders of San Diego's Stone Brewing.** (Studio Schulz)

**Below: The high ABV/high IBU lineup at Stone includes the flagship Arrogant Bastard, Double Bastard, IPA, Ruination and Levitation.** (Studio Schulz)

Rattner, and Cramer's cousin, Karl Strauss, who had spent 44 years with Pabst. A decade later, they expanded, with a large production brewery in Pacific Beach and other brewery restaurants across San Diego County and beyond.

They now operate restaurants at the international airports of both San Diego and Los Angeles as well as at the Universal Studios "Citywalk" entertainment district.

The Karl Strauss lineup of regulars includes 3.3 percent ABV Endless Summer Light, 4.2 percent Karl Strauss Amber, 5.5 percent Pintail Pale Ale, 5.8 percent Red Trolley Ale, 6.5 percent Tower 10 IPA, and 5.1 percent Windansea Wheat. The latter is named for Windansea (*not* Windandsea) Beach, a legendary San Diego County surfing location.

Stone Brewing of Escondido in San Diego County owns a box seat on the leading edge of the "edgy" category within the genre of extreme brewing. Founded in 1996, Stone has grown steadily,

occupying a seat among the top 20 brewing companies in the United States. In 2011, the readers of *Beer Advocate* magazine named it as the "All Time Top Brewery on Planet Earth." Typical of many California brewers in the twenty-first century, Stone's beers are highly to extremely hopped, frequently in double digit BV percentage, and often aged in oak. Stone is perhaps best known for their 7.2 percent ABV Arrogant Bastard Ale, which is marketed with the arrogant slogan, "Hated by many. Loved by few. You're not worthy." Other Stone products include 5.4 percent Stone Pale Ale, 4.4 percent Stone Levitation Ale, 7.2 percent Oaked Arrogant Bastard Ale, and 7.7 percent Stone Ruination IPA.

Other important San Diego County breweries include AleSmith Brewing, started in 1995 by Skip Virgilio and Ted Newcomb, and now owned by brewmaster, Peter Zien; Ballast Point Brewing, founded in 1996 by Jack White

**Above: Chris Cramer and Matt Rattner, founders (with Karl Strauss) of San Diego's Karl Strauss Brewing. (Karl Strauss)**

**Below: The Karl Strauss portfolio includes Red Trolley Ale, Tower 20 IPA, and Windansea Wheat. (Karl Strauss)**

264

Above: Half Moon Bay Brewing in Princeton, California. (Bill Yenne photo)

Above right: A vintage coaster from Half Moon Bay Brewing. (Author's collection)

Below: Double Barrel Ale and Union Jack IPA from Firestone Walker Brewing. (Firestone Walker)

and Yuseff Cherney; Green Flash Brewing, started in 2002 by Mike and Lisa Hinkley; and Lost Abbey Brewing, founded in 2006 in a brewery that was previously operated by Stone Brewing.

On California's Central Coast, Firestone Walker Brewing was started in 1996 by brothers-in-law David Walker and Adam Firestone, the son of Brooks Firestone, the owner of Firestone Vineyard. Located in Paso Robles, their brewery "was inspired by the old Burton Union system, once a staple of British beer making. The Burton Union was developed around 1840 as English tastes shifted from London porters to pale ales crafted in Burton-upon-Trent." Firestone Walker is thought to be the only brewery in the United States to currently employ the "double barrel" union brewing method. Their

products include 5 percent ABV Double Barrel Ale and 12 percent Double Double Barrel Ale as well as the critically acclaimed 12.5 percent Abacus barleywine, 9.5 percent Double Jack, 7.5 percent Union Jack IPA, and 8.3 percent Wooky Jack. Their 13 percent Parabola imperial stout ranks in the *Beer Advocate* top ten worldwide.

Farther up the coast, a favorite of this author is the Half Moon Bay Brewing brewpub in Princeton-By-The-Sea, an unincorporated village just north of the town of Half Moon Bay, which is adjacent to the well-known Mavericks surfing location. Founded in 2000, they have always offered 4.8 percent ABV Mavericks Amber Ale as their flagship beer, and today, the entire lineup bears the Mavericks name. Other beers include 4.8 percent Mavericks Big Break Golden Ale, and 4 percent Mavericks Bootlegger's Brown Ale.

In a list headed by the city's well-established Anchor Brewing, San Francisco is home to well over a dozen brewpubs and micro-breweries of varying sizes and has seen at least that many come and go since Allan Paul started San Francisco's first brewery since Prohibition back at the dawn of craft brewing. Some of the early names included the Twenty Tank Brewery. Gordon Biersch opened one of their earliest brewery restaurants here in 1992 but closed it in 2012. This left Thirsty Bear Brewing, a brewery restaurant specializing in Spanish cuisine that opened in 1996, as the city's oldest brewery restaurant.

The Beach Chalet Brewery was established in 1997 by Gar and Lara Truppelli and Timon Malloy in a 1925 Spanish Colonial Revival–style restaurant building overlooking the Pacific Ocean at the western end of San Francisco's Golden Gate Park.

Also in 1997, on Haight Street at the opposite end of Golden Gate Park, a passionate university-trained brewer named Dave McLean opened his Magnolia Gastropub & Brewery. McLean opened a second gastropub in the Dogpatch neighborhood in 2014. He writes that his pub "marries a Slow Food-like respect for tradition with a Grateful Dead-esque approach to inspired creativity." The choices at McLean's gastro-pubs vary widely, but 4.8 percent

Kalifornia Kölsch, 4.8 percent Cole Porter, 7.2 percent Proving Ground IPA, and 10.2 percent Magnolia Old Thunderpussy Barleywine are typically offered.

Speakeasy Ales & Lagers was founded in 1997, not as a brewpub, as were so many San Francisco breweries of its era, but as a production brewery. Nevertheless, they do have a tap-room and host frequent Friday night parties. Speakeasy beers, branded with the iconography of the Prohibition era, are widely available in California and other states. Their beers include 6.1 percent ABV Prohibition Ale, 6.5 percent Big Daddy IPA, 9.5 per-cent Double Daddy Double IPA, 5.5 percent Scarlett Red Rye, and 8 percent Butchertown Black Ale.

Across town, another San Francisco brewery celebrates the constitutional amendment that *ended* Prohibition. The 21st

**Above: Shaun O'Sullivan of the 21st Amendment Brewery and Richard Brewer-Hay of the Elizabeth Street Brewery. (Bill Yenne photo)**

**Below: Dave McLean's Magnolia Gastropub & Brewery on San Francisco's Haight Street. (Bill Yenne photo)**

Amendment Brewery, best known locally as "21A," was started in 2000 by Nico Freccia and Shaun O'Sullivan, a veteran of Berkeley's Triple Rock Brewery. The brewpub is located only two blocks from the major league ballpark that is home to the San Francisco Giants, though brewing and canning for outside distribution takes place at Cold Spring Brewing in Cold Spring, Minnesota. O'Sullivan's popular portfolio includes 8 percent ABV 21st Amendment IPA, 6.8 percent Back In Black black ale, and the popular 4.9 percent Hell Or High Watermelon Wheat Beer.

Also available recently was 8.5 percent Imperial Jack ESB, created by Richard Brewer-Hay at San Francisco's Elizabeth Street

Brewery. The story of the latter beer is one of those remarkable tales from brewing lore that begs retelling. Richard had been brewing an ESB named "Grandpa Jack" in honor of his late grandfather, Jack Newbould, when he and O'Sullivan decided to collaborate on a stronger version at the 21A brewery, renaming it "Imperial Jack." As they explain, "in 2010, we entered it in the international World Beer Cup competition, and it won a gold medal in the Other Strong Ale category." They later traveled to London, where they brewed Imperial Jack with Angelo Scarnera, the head brewer at Brew Wharf. They describe this as "an amazing event for both of us, the chance to brew this beer in London. The idea of bringing the Imperial Jack recipe back to England where Richard's grandfather spent his life was a real emotional journey for Richard, one that will be cherished for a lifetime."

North of San Francisco, in Sonoma County, there is a concentration of some of the most well-known breweries in the United States.

Bear Republic Brewing was founded in 1995 by brewmaster Richard G. Norgrove and his father, Richard R. Norgrove, who serves as the company's CEO and master of sales and marketing. Many of the brewery's products bear references to Richard G.'s ongoing second life as a stock car

driver, which has become part of Bear Republic's image. When this author visited him at the brewery, he pointed out a car door, battle damaged in a recent race, that was going to be displayed on the wall of one of the brewery's retail accounts in New York City. Bear Republic operates a production brewery in Calistoga as well as their original brewery and restaurant in Healdsburg, where smaller, experimental brews are carried

These pages: A look inside Bear Republic Brewing of Cloverdale, California. Above is Richard G. Norgrove, the co-owner and brewmaster; while metal meets the oak in the Bear Republic cellar on the opposite top. (Bill Yenne photos, both)

Opposite: The lineup includes the flagship Racer 5 IPA, Red Rocket Ale, Hop Rod Rye IPA, Big Bear Black Stout and Café Racer 15, with IBUs in excess of 100. (Bear Republic)

Above: Lagunitas brewmaster Jeremy Marshall in the Petaluma brewhouse with Mary Bauer, head brewer at the Lagunitas Chicago brewery. (Bill Yenne photo)

Below: The flagship Lagunitas IPA and the seasonal Little Sumpin' Wild. (Lagunitas)

Below right: The Little Sumpin' spokesmodel on the brew kettle window. (Bill Yenne photo)

yeast species indigenous to Sonoma County.

Lagunitas Brewing was started in 1993 by homebrewer Tony Magee in Lagunitas, but the popularity of the beers dictated a move to a location in nearby Petaluma with freeway access for the trucks. One of the fastest-growing craft breweries in the United States, their output quadrupled between 2004 and 2010, catapulting it into the top ten of American brewing companies. With capacity topped out in Petaluma, a second brewery in Chicago, with a 600,000 barrel capacity, was opened in 2014. The best-selling IPA in a state where IPA is the de facto national beer style is the brewery's 6.2 percent Lagunitas IPA, the flagship beer. This is complemented by 8.2 percent Maximus IPA, a "bigger" IPA with 70 IBUs, and 8 percent Hop Stoopid, a double IPA with over 100 IBUs that is made with hop extract and oils rather

out. The flagship beer is 7 percent ABV Racer 5 IPA. Other products include 6.8 percent Red Rocket Ale, "a bastardized Scottish style red ale that traces its origins to our homebrew roots," as well as 8 percent Hop Rod Rye IPA, 8.1 percent Big Bear Black Stout and 9.75 percent Café Racer 15, a double IPA with IBUs in excess of 100. In Healdsburg, cask-contained ales are being aged, and experimental lambic-type beers are being made using spontaneously-fermenting *Brettanomyces*

than hop flowers. Also produced outside the IPA family are 6 percent Lagunitas Pils, 9.99 percent Brown Shugga, and 7.5 percent Little Sumpin' Sumpin' Ale, with "Millie," the iconic, but imaginary, spokesmodel on the label. A higher gravity seasonal variation on the latter is 8.8 percent Little Sumpin' Wild, made with Westmalle Trappist yeast.

At Russian River Brewing, two people who grew up in the California Central Valley wine industry bought a brewery from a winery and came to brew what are recognized by critics and consumers as some of the world's greatest beers. Vinnie Cilurzo and his wife Natalie came to Sonoma County in 1997, when Vinnie left his Blind Pig Brewing Company in Temecula to come to work at Russian River when it was owned by Korbel Champagne Cellars. In 2003, Korbel left the beer business and sold the brewery to the Cilurzos. Today, they maintain a brewpub in downtown Santa Rosa, as well as a production brewery where they produce their beer for a growing circle of retail accounts in the West and as far east as Pennsylvania.

Accounting for more than half of their production, the flagship brand is 8 percent ABV Pliny the Elder, a double IPA named for the ancient Roman

historian who discussed hops in his early encyclopedia of natural history. Hops are important to Vinnie Cilurzo. They play a key role in finely balancing everything that he brews, even the handful of beers on his list that are less than 6 percent

**Above: Ranks of fermenters that are part of the 600,000-barrel capacity at Lagunitas Brewing in Petaluma, California. Annual output here quadrupled in the first decade of the twenty-first century. (Bill Yenne photo)**

Above: Inside the brewhouse at the Russian River Brewing Company's main production site on the outskirts of Santa Rosa, California. (Bill Yenne photo)

Below: Major Russian River brands include the flagship Pliny the Elder, as well as the barrel-aged Sanctification, Supplication, and Temptation. (Russian River)

nephew. Pliny the Younger is a 10.5 percent triple IPA which, along with its uncle, shares a place on most lists among the best half dozen or so beers in the world. While the Elder is widely available, the Younger is so rare as to have global cult following. Released for just two weeks each February, it is available only at the Brewpub and at select accounts in California. The lines rival those at a Hollywood premier. A 2013 report from the Sonoma County Economic Development Board devoted three pages solely to the $2.36 million economic impact of the two-week release of Pliny the Younger. Other Russian River beers include 6.1 recent Blind Pig IPA, 7 percent Damnation golden ale, and a long list of barrel-aged beers fermented with *Brettanomyces*, including 5.5 percent Beatification and 7 percent Supplication, aged 12 months with sour cherries.

ABV. One might say that he has forgotten more about hops than most of us will ever know, except that one doubts he has forgotten anything.

The nephew of the beer named for Pliny the Elder is named for the historian's actual

Across the Pacific in Hawaii, the American brewing tradition goes back to the Honolulu

Brewing Company, started in 1898, and the Hawaii Brewing Company (also in Honolulu) that was formed in 1934. The latter became famous for its Primo brand that dominated the Hawaiian market until Schlitz bought the company in 1964. Primo faded quickly when Schlitz moved production to the mainland in the 1970s, and the brand disappeared for decades before being revived in 2007. Breweries operating in Hawaii today include the Aloha Beer Company in Honolulu, Hawaii Nui Brewing in Hilo, and Maui Brewing in Lahaina.

The largest and best known craft brewing company in Hawaii is Kona Brewing on the Big Island of Hawaii. Started in 1994 by the father and son team of Cameron Healy and Spoon Khalsa, Kona's first bottling was in February 1995. In 2010, Kona joined with Widmer and Redhook as part of the Craft Brew Alliance, which gave Kona access to their production facilities to brew its beers for the mainland market. Kona's

**Above: Natalie and Vinnie Cilurzo, owners of the Russian River Brewing Company, enjoy glasses of Pliny the Elder while his cousins age in the background. The smaller photo shows the racks of aging barrels.**
(Bill Yenne photos)

Above and below: Kona Brewing Company, based in Kailua-Kona on Hawaii's Big Island, is the largest brewery in the state. Longboard and Pipeline are two of their leading brands. (Kona)

Above right: Inside Kauai Island's brewpub. (Kauai Island)

flagship beers are 4.6 percent ABV Longboard Island Lager, 4.4 percent Big Wave Golden Ale, and 5.9 percent Fire Rock Pale Ale. Kona also produces seasonal brands, including 5.5 percent Koko Brown Ale, and beers such 5.6 percent Lavaman Red, which are available only in Hawaii.

At the opposite end of the Hawaiian archipelago, in Port Allen on the island of Kauai, is "The World's Westernmost Brewery." As they say at Kauai Island Brewing, their beer is the "last beer before tomorrow." Indeed,

for the beer traveler, this is the end of the line. There are no breweries on earth between here and the International Dateline. The previous westernmost brewery was Waimea Brewing, was about a half-dozen miles farther west. When the lease was up at this location in 2011, owners Bret and Janice Larson, along with their brewmaster Dave Curry and his wife Christina, decided to come together and become co-owners of Kauai Island, which opened as a brewpub in 2012.

The beers which Dave Curry has brewed for the travelers who come here to watch the sunset at the last frontier of the world of beer include 5.4 percent ABV Na Pali Pale Ale, 4.3 percent Pakala Porter, 3.8 percent South Pacific

Brown, 4.3 percent Leilani Light, 6.3 percent Cane Fire Red, and 4.8 percent Wai'ale'ale Ale, a beer with a tongue-in-cheek-twisting name that pays homage to Kauai's Mount Wai'ale'ale, the alleged wettest place on earth. Travelers will enjoy the 5.5 percent Captain Cook's IPA, named for a fellow traveler from the tall ship era who "discovered" Hawaii. Indeed, his first landfall

in Hawaii was only eight miles from the brewery's location. As Curry writes, "Created after the hoppy ales brewed in the 1700s for the long journey from England to Hawaii, this Ale is one of our hoppiest beers ever [67 IBUs] using 15 pounds of hops for 155 gallons of beer." Grab one and join me outside for the sunset, there's a tall ship on the horizon.

**Above: Bret Larson and Dave Curry of the Kauai Island Brewing Company enjoy a round at their pub in Port Allen on the island's south shore. It is the westernmost brewery in the world. (Kauai Island)**

**Above left: The Kauai Island brewhouse. (Bill Yenne photo)**

**Below: This selection from the Kauai Island portfolio includes Captain Cook's IPA, Leilani Light, Napali Pale Ale, and Wai'ale'ale Ale. (Kauai Island)**

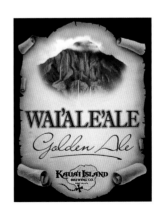

# An Overview of Beer Styles

Some lists of beer styles run to more than 100 entries, but rather than merely listing them, this overview attempts to group and classify them. As within any system of taxonomy, such as the classification of species in biology, we organize the styles into groups with similar characteristics. Some beers are classified by color, some by their malt bill (the proportions of malted grains used), some by their country of origin, and some by the species of yeast used in fermentation. The latter is how brewers typically begin to classify beers. In the following descriptions, generalizations are made and it should be understood that there are exceptions to every rule.

The vast majority of the beers in the world are lagers. They are fermented with *Saccharomyces pastorianus*, formerly called *Saccharomyces carlsbergensis*, and known generically as lager yeast. It is a bottom-fermenting yeast because it does the work of converting starch and sugar into alcohol and carbon dioxide at the bottom of the fermenting vessel. It ferments at cold, near-freezing temperatures and came into widespread use in the nineteenth century after it became possible to easily transport ice and with the advent of artificial refrigeration. The term "lager" means to store for a long time, and lagers are indeed "lagered" for extended periods, often around a month or longer. Lagers are typically described as "crisp" or "clear." They range in color from a very light yellow to black. Most examples are on the lighter end of the color spectrum.

Most of the remaining beer in the world is made with top-fermenting *Saccharomyces cerevisiae*, known as ale yeast. Top-fermenting yeast works at warmer temperatures than bottom-fermenting yeast, and fermentation is usually deemed complete in about half the time. Ales are typically described as "rich" or "full-bodied." They range in color from a very light yellow to black. Most examples are in the amber part of the spectrum. Other major classifications of beers using ale yeast are porters and stouts.

A tiny fraction of the beer made in the world uses other yeasts or microorganisms. Examples of these are the species of yeasts within the genus *Brettanomyces*, known to brewers as *"brett."* Once used mainly in Belgium, they are also increasingly popular among craft brewers, especially in the United States.

## The Lightest, or Pale, Lagers

In these beers, pale color and clarity is the desired feature. They are designed to be consumed cold as a thirst quencher and are most likely to be "flavor-neutral," because it is hard to maintain consistency in beers with assertive flavors. These beers are often called adjunct lagers because the light color, body, and flavor is achieved by using adjunct grains, such as rice or corn, in addition to barley malt. Most of the mass-market beers throughout the world, especially in the United States, are in this category. The use of adjuncts also means they are cheaper to produce, making them ideal for companies wishing to compete for shelf space on the basis of price.

Included in this class are "light beers," which are lagers low in carbohydrates and calories. Other types, such as "dry beers" and "ice beers," are indistinguishable subclassifications of the same category, and were created mainly for the purposes of competing for shelf space.

Most of the lagers in this category have between 4 and 5 percent ABV. Exceptions are non-alcoholic beers, which are not literally without alcohol, but are usually around 0.5 percent ABV. On the other end, "malt liquor" is a higher-alcohol exception. This term does not describe a separate beer style, but rather it is a term coined by statute to describe higher-alcohol beer for the purpose of controlling its sale in certain jurisdictions. Depending on the jurisdiction, malt liquor is usually between 5.5 and 8 percent ABV. Contrary to the implication of the name, it has virtually no malt char-

acter. Contrary to the statutory intentions, it is typically sold in 40-ounce containers and marketed in poor neighborhoods.

## Pilsners and International Lagers

These lagers are similar to pale lagers in appearance, though more straw-colored than pale yellow, and are usually made only with malted barley. Therefore, they have more flavor and aroma and a fuller body. They are typically between 4 and 6 percent ABV. These lagers are very much like the original commercial lagers that emerged in the 1840s from the vicinity of Munich in Bavaria and Pilsen in what is now the Czech Republic.

The style that originated in Pilsen is still made there, but it is copied throughout the world, where it is known as pilsner (sometimes spelled pilsener) or simply called pils. It is defined in part by the use of very floral Saaz hops, which are grown in the region. In Germany and elsewhere, hops

other than Saaz may be used in pilsners, but the overall feel is similar.

Similar beers that match the same general definition, but which often use more bitter, less floral hops, include the all-barley malt lagers made by craft breweries, especially in North America, as well as the German styles helles (meaning bright) and export, which originated as a style in the German city of Dortmund. International lagers and Euro lagers are also in this category, but these terms are more general and function as an umbrella for straw-colored barley malt lagers that are in wide distribution.

## Amber Lager

These beers are similar in many ways to the Pilsners and International lagers, except that the malt is roasted a bit darker. The style may have its origins with the Vienna of the early nineteenth century, which was somewhat darker than its Pilsen and Munich cousins. They are typically between 4 and 6 percent ABV.

## Märzen

A specific genre of lager, märzen may be straw colored or amber. Named for the German word for the month of March (März), it originated as a beer brewed in that month, then aged or lagered in ice caves through the summer, and served during Oktoberfest. Beers which are called Oktoberfest beers are märzens. They are typically between 4 and 6 percent ABV, but some may go a bit higher.

## Dark and Black Lagers

It may be said that lagers that run darker than amber can be as dark and full-bodied as top-fermented porters and stouts, but have a lighter feel, though this is a generalization. Most of these beers have a pleasing ruby-red color when held up to a light source.

There are dark or black lagers made in many countries, and some of these use the stylistic names used in Germany, such as dunkel (dark) and schwarzbier (black beer).

In general, they are similar to lighter lagers in many characteristics such as the types of hops used, but the malt is obviously roasted much darker. They are typically between 4 and 6 percent ABV, but some may go a bit higher.

## Rauchbier

A rare style of lager that originated in Bamberg in Bavaria, rauchbier is literally "smoked beer." The malted grain used in the brewing of rauchbier is roasted over the open flame of a beechwood fire. This imparts a distinctive smoky flavor to what looks visually like a dark lager. Since the 1990s, brewers elsewhere have made smoked beers in a similar manner, some fermented with lager years, others with ale yeast.

## Bocks, Doppelbocks, and High-Alcohol Lagers

Bock beers are similar to märzens in many ways. They are robust and full-bodied, usually amber in color, have an ABV content slightly greater than average lager, and they were traditionally brewed at a specific time of year. As märzens were brewed in the spring for autumn consumption, bock beers were brewed around December for consumption in the spring. Pictures of goats are typically associated with bock beers, but there is no definitive explanation of this tradition.

Doppelbocks, meaning "double bocks," are a richer, more full-bodied variant of bock beers, with much more maltiness. They are also usually higher, though not *double*, in ABV. Bocks range between 5 and 7 percent, while doppel-

bocks, with some exceptions, range between 6 and 9 percent.

Another higher-alcohol German beer style is Eisbock, meaning "ice bock." They are essentially bock beers in which some of the water is extracted by fractional freezing. This results in beers that can be as strong as the 15 percent ABV range. The process is similar to that used for North American "ice beers," although with these, water is added back in to reduce the ABV.

American malt liquor is another form of high-alcohol lager, usually less than 8 percent ABV, but it is flavor neutral and marketed only for its alcohol content.

## English Ale

Generally, though not universally, English ale is a full-bodied beer amber in color and in the range of 3 to 6 percent ABV. A defining feature is that it is made with English hop varieties and is therefore more bitter than German-inspired lagers. Naturally, ale is made with top-fermenting yeast.

In Britain, ale is considered to be "real ale" if it is served at the pub from unpressurized casks. Americans unfamiliar with this practice (although many American brewpubs now offer cask selections) often find cask ale to seem "flat," or lacking in carbonation. Ale served this way has a different taste than essentially the same beer from a bottle or pressurized keg. In the pub, one often describes cask ale as bitter or mild, depending on hoppiness. Another variation is extra special bitter (ESB), which is basically similar to the above but more assertive in malt and hop content, and perhaps as strong as 7 percent ABV or so.

The definitive bottled ale is pale ale, which is not light yellow, like a pale lager, but amber. As a style, it was defined in the eighteenth century by the great brewing companies in Burton-upon-Trent, but today it is the staple not only of British brewers, but of nearly every craft brewery in North America.

Branching out from the above generalizations in terms of color brings us to dark ale and brown ale, which are exactly as their names suggest and much fuller in malt flavor.

Branching out from the above generalizations in terms of hop content brings us to India Pale Ale (IPA), created in the eighteenth century as a pale ale variant high in alcohol and high in bittering hops, which were essential to preserve the beer when exported on long sea voyages to distant parts of the British Empire, such as India.

## American Ale

American ale styles are generally analogous to English ale styles, although cask-conditioned ales are definitely less common. In the late twentieth century, pale ale in the range of 4 to 7 percent ABV became the definitive American craft beer, although it was often referred to as amber ale or red ale. At the turn of the twenty-first century, IPA became the popular lead beer style of many craft brewers in North America, who created their own stronger variants, double and triple IPAs. North American brewers also embraced porters and stouts originally based on British prototypes.

## Scottish Ale

Generally similar to darker English ales, Scottish ale tends to be fuller flavored and sweeter, with malt characteristics overpowering the hops. A variant called wee heavy ranges up into the 8 to 10 percent ABV range.

## German Alt

In Germany, top-fermented beer was the norm until the early nineteenth century, after which lager took over almost completely. It still survives proudly as a regional tradition. In Düsseldorf, they brew altbier, meaning beer of the "old" style, which is similar to an English brown ale. Cologne (Köln) is home to Kölsch, which is light in color, like a pilsner, and mild in terms of hop and malt character. Both styles average in the 4 to 7 percent ABV range. They remain rare outside their home turf, but some American craft brewers have offered these styles.

## Belgian Ale

Amid a brewing tradition so complex, it is hard to generalize about Belgian ales. One might say that they are more complex and more delicate, that the hops used are less bitter than in England, and that the yeast strains used impart a distinctively recognizable flavor. They exist in the same color range as other beers, from the straw-colored golden ales, topped with a large, dense, long-lasting white head, to very dark brown. In Belgium, beer of any color or shade that is roughly between 6 and 9 percent ABV is frequently called a "dubbel," with a "tripel" being a beer with an ABV between 9 and 12 percent. In recent years, some brewers have used the term "quadrupel" to identify beers of yet higher ABV.

## Oud Bruin

A dark ale that originated in Flanders and the Netherlands, oud bruin (old brown) is still popular and widely produced in the Netherlands, where the brewing industry is more heavily dominated by lagers than it is in Belgium. It is also called Dutch brown and ranges between 5 and 8 percent ABV. Oud bruin is rich in malt, light on hop flavor, and is usually aged for several months, often as much as a year.

## Porter

Porter is a very dark, extremely full-bodied top-fermented beer. Born in London in 1722, it was originally a favorite of Covent Garden market porters. It was popular in Britain and Ireland into the nineteenth century, but faded from the mainstream until reintroduced by craft brewers in the late twentieth century. It ranges from about 4 to 7 percent ABV. A late eighteenth-century variant was Baltic porter, with up to around 20 percent ABV, that was developed in Britain for export to Baltic countries, including Russia. It is still a popular style in the region and still produced locally. Typical modern Baltic porters are in the range of 7 percent ABV, and are actually black lagers, not true porters.

## Stout

From porter, there evolved the blacker, richer "stout porter," which came to be called stout, in which malt has been roasted to nearly black. In the early nineteenth century, Ireland became a leading producer, and variants of Guinness's family of stouts remain definitive of Irish stout. In England there were and are many varieties of stout, including

sweet and milk stouts. The addition of oats to the mash while brewing produces the especially smooth oatmeal stout. Since the craft brewing revolution in the 1980s, stouts have been a common addition to the portfolios of brewers from England to North America. In the late eighteenth century, the same British brewers who invented Baltic porter created the very strong Russian imperial stout. Enjoying a revival among craft brewers, modern imperial stout is often rated at 12 percent ABV or above.

## Barleywine

Not really wine, of course, barleywines are generically the strongest of strong ales, characterized by being sweeter and less hoppy than most ales and much stronger. Barleywine is usually in the range of 7 to 12 percent ABV. Having originated in England, it has been copied by many craft brewers in North America. Like wine, it ages well in the cellar because of the alcohol.

## Wheat or "White" Beer

As the name implies, wheat beer is made using a high proportion (50 percent or more) of wheat malt with the barley malt, with the result being a light-colored, though not literally "white," beer. In German, the terms weissbier (white beer)

and weizen (wheat) are both used to describe the beer. One of the most popular wheat beer styles, with both German and American brewers and their customers, is hefeweizen, an unfiltered golden wheat beer, which is translucent because yeast (hefe) remains in suspension. Like ale, German wheat beers are generally top fermented, but American brewers also produce other varieties of wheat ales as well as wheat lagers.

Belgium also has an indigenous unfiltered wheat beer. In Flemish, a white beer is a witbier and in French it is called bière blanche. The Belgian white has the cloudy appearance of a German hefeweizen but has a quite distinct flavor. The Belgians flavor their "whites" with coriander and sometimes orange peel. Other spices are also used by some brewers, but coriander is definitive of the style.

Despite the common naming convention, not all wheat beers are white. For example, many German weissbier brewers also brew dunkelweizen, a dark, opaque unfiltered wheat beer.

## Lambic

The lambics are a unique wheat beer variety indigenous to Belgium. They have been made for generations in the Senne Valley, using spontaneously fermented with strains of wild *Brettanomyces* yeast native only to this area. An inherently sour beer, lambic was traditionally flavored with tart wild cherries and named kriek, after the Flemish word for the dark-red morello cherry. By the twentieth century, kriek lambics were being joined by framboise (or frambozen) lambics flavored with raspberries and cassis lambics flavored with black currents. By the 1990s, brewers were marketing lambics flavored with apples, apricots, grapes, peaches, plums, strawberries, bananas, and even pineapples. Lambics made with a mixture of flavors and aged for several years are called gueuze. Lambics that are flavored with sugar, molasses, or other beer styles are called faro.

## Farmhouse Ale

Homebrewing for personal or local consumption is a tradition that has existed from the beginning of brewing. In the nineteenth and twentieth centuries small breweries serving villages or small rural areas either evolved into large breweries or disappeared. One area where the farmhouse tradition has survived and flourished is in the Nord-Pas-de-Calais region of France and in adjacent rural parts of Belgium's Wallonia. Most of the beers were traditionally produced in the winter or spring for later consumption through the year. They span the range of beer styles, although most are top-fermented and unfiltered.

In France, they are called bière de garde, meaning "beer for keeping," as they are usually aged for several months. The similar beers from Belgium are called saison, meaning season, because they were once brewed in the winter for consumption by seasonal workers known as "saisonniers." In the late twentieth century, some examples of both were commercialized, and they have become popular in the twenty-first century. Originally, they may have been less than 4 percent ABV, but modern commercial examples are twice this number.

## Chibuku

In southern Africa, Chibuku is a widely distributed black "opaque beer," that is also known locally as "scud" or "shake shake." It is brewed with sorghum and maize rather than barley malt and ranges between 3.5 and 6 percent ABV. It is produced commercially by many large brewing companies, including SABMiller. Chibuku is traditionally sold in paper cartons similar to milk cartons, though it may also be found in plastic bottles.

## Chicha

Having originated in the Inca Empire in pre-Columbian times, chicha is a traditional fermented beverage based on corn or cassava, but sometimes made from fruit. It is still widely produced

and enjoyed throughout Central and South America.

## Japanese Rice Beer

Japanese rice beer is not the same as sake. Sake, often called rice wine in the West, is made with rice, but brewed like beer. It is then filtered and fortified with distilled spirits and served at about 18 percent ABV. Rice beer is a term for beer brewed with a high proportion of rice. If the beer is mostly rice, it is called happoshu, which is between 4 and 7 percent ABV. Very popular in Japan, happoshu is made by most large Japanese brewing companies as part of their product line.

## Kvass

Virtually unknown in the West, kvass is traditional beer style indigenous to Russia and Eastern Europe, which, like happoshu in Japan, is widely popular and is brewed by large commercial brewing companies. The exact definition varies, because traditionally kvass could be made with virtually any fermentable starch, especially rye bread, using baking yeast. It may or may not be flavored with fresh or dried fruit. With only around 1 percent ABV, it is available by the glass from street vendors wheeling large tanks on carts. Commercial kvass is bottled.

# An Overview of Beer Glassware

Above: The new Guinness glass has sculpted sides, but retains the classic tulip shape. (Diageo)

Below: Classic Guinness tulip glassware at a pub. (Bill Yenne)

Glassware is an important and integral part of the enjoyment and the presentation of beer. A good beer appeals to a person's sense of taste. A good glass helps the beer appeal to other senses, augmenting the experience visually, and helping to focus the aroma, further enhancing the encounter with the beer. While most serious beer enthusiasts pride themselves on having and using collections of glassware in various sizes and shapes, this chapter is devoted to glassware which one might commonly encounter in the average bar, pub or bierstube.

In the United States and the United Kingdom, the most common beer glass is the standard and ubiquitous, straight-sided, slightly conical *pint glass*. It originated in the United Kingdom with a volume of one imperial pint (568 milliliters), and since the craft brewing renaissance of the 1980s, it has been almost universally adopted by bars and restaurants throughout the United States. Here, these glasses have a volume of one American pint, or 16 fluid ounces (473 milliliters). In both the United States and the United Kingdom, half-pint glasses were once common, but in the United States today, they are rarely seen in bars.

There are many variations on the straight-sided pint glass that add a slight curvature to the profile. These alternate shapes help to trap and focus the aroma and maintain the head after pouring. They include the *nonic* glass, which is like the straight-sided glass, but with a slight bulge a short distance below the rim, and the *tulip* pint glass made famous by Guinness.

Even as there has been a proliferation of beer styles in the United States, the variety of glassware has been replaced by the uniformity of the straight-sided pint glass. The *schooner*, a chalice-shaped beer glass, was once widely seen in the United States, though it gradually faded from common use in the late twentieth century. The same is true for mugs with handles, which are seen at home, though they are not common in bars. The *dimpled* pint mug is still common in the United Kingdom, though rarely found in the United States. Straight-sided mugs are seen the United States, but are not common.

Pewter or wooden mugs and tankards, with or without hinged lids, which were common in the eighteenth and nineteenth century, are seen as display items, but are rarely used for drinking.

In Australia, a *schooner* is a measurement of beer usually equal to 15 fluid ounces or 425 milliliters. Traditionally, Australians order their

beer by diverse names based on the measure, but the names can vary by location. The smallest is the *pony*, which is roughly five fluid ounces or 140 milliliters. In some places, the 10-ounce pour is called a *middy*, though elsewhere it is called a *pot* or a *ten*. At seven ounces, a *glass* is between a pony and a middy.

In Germany, Austria, and neighboring countries, the standard glass, the equivalent to the Anglo-American pint glass, is the *willibecher*, which has a slightly curved side profile. In German, the root word becher translates as *cup*, and a willibecher is roughly translated as a *tumbler*. These glasses are generally found in a half-liter (500 milliliter) or 330 milliliter size, though other sizes are also common.

In Germany and Austria, specific beer styles have glasses that

Above: Melody Daversa, the marketing, PR, and ad manager at the Karl Strauss Brewery in San Diego, sips from the standard straight-sided, Anglo-American pint glass. (Karl Strauss)

Below: The "willibecher" is the standard beer glass in Germany, Austria and adjacent central Europe. (Bräu Union)

**Above: Edelweiss hefeweizen and dunkelweiss, both unfiltered wheat beers, in classic weiss or weizen glasses; Paulaner Pils in the classic pilsner glass; Summit Pilsner in a wide-bottomed pilsner glass; and Gaffel Kölsch in a straight-sided *stange*.** (Courtesy of the breweries)

**Below: Enjoying hefeweizen from the proper glass** (Bill Yenne photo); **and having fun in the biergärten at the Munich Hofbräuhaus with the mighty one-liter *maßkrug*.** (Hofbräuhaus photo)

are specifically shaped. One of the most common and most uniform is the glass used for wheat beer (weissbier, weizenbier or hefeweizen). This glass has been adopted in the United States and elsewhere for serving hefeweizen, which is now a staple of many brewers outside of the German-speaking world.

While the *hefeweizen glass* conforms to a specific appearance, the *pilsner glass* is found in many

variations. The only thing that most pilsner glasses have in common is that they are tall and narrow. Some may be conical, with a slightly wider top, while others have no variation in width. In German, the latter are called a *stange*, meaning *rod* or *stick*. Pilsner glasses, both of the conical and stange variety are also found in sizes smaller than a half liter, such as 200 or 330 milliliters. Ironically, at Pilsner Urquell in the

Czech city of Pilsen, home of the "original pilsner," the beer is served in straight-sided, handled mugs (see page 77).

In Europe, and to a lesser extent elsewhere, beer may also be served in stemware. Some pilsner glasses have short stems, and many specific beer brands are served in their own designated snifters, flutes or goblets. In German the goblet is known as a *pokal*, and this is a com-mon term in the lexicon of German beer glasses.

In much of the world, the standard pour of beer in a bar, pub or bierstube is in the range of a pint or a half liter. In Munich, and to a lesser extent elsewhere in Germany, where beer drinking is a serious sport, the standard glassware is the one-liter, handled *maß*

Trumer's Berkeley brewmaster, Lars Larson, presents a portion of Trumer Pils in one of Trumer's *schlanke stange*, or *slender rod* glasses. In 1997, the glass was declared to be "the most beautiful beer glass in the world," and featured in an exhibit at the Museum of Modern Art in New York.

As Larson explained to the author, "it's an elegant design, lighter to hold, and it feels good in your hand. You can see the clarity and color of the beer very well. The straight sides are important as a design element. Straight-walled and narrow at the top also helps with the lacing of the beer, and helps keep the fragrance in, so that you get more of the hop aroma."
(Bill Yenne photo)

**Top left: Examples of beer in stemware, including Köstritzer Schwarzbier; Ksiazece Czerwony Lager from Piwowarska; and Brackie Mastne from Zywiec. (Courtesy of the breweries)**

**Left: A classic dimpled pint mug on the window sill at an English pub. (Greene King)**

**Above: A variety of Belgian beer glasses: Stella Artois stemware; Hoegaarden's tumbler; the Lindemans lambic flute; Boon's lambic goblet; and the straight-sided Palm tumbler.** (Courtesy of the breweries)

**Inset: The delicate and unusual Kwak glassware.** (Bill Yenne)

**Below: Intricate lacing on a chalice of beer at the Brouwerij de Koningshoeven. Trappist glasses are typically chalice-shaped.** (Koningshoeven)

or *maßkrug*. When experienced for the first time by outsiders, the sight a vast beer hall full of people drinking from maßkrugs can be hugely astonishing. Less common today than in the mid-twentieth century and before, is the one-liter stoneware *stein* and the half-liter stoneware *humpen stein*. The word stein literally translates as *stone* (*stoneware* is *steingut*), but any German drinking vessel with a handle is typically called a stein in the Anglo-American vernacular. Many have hinged metal lids, the practical purpose of which was to keep falling leaves and other debris out of the beer in outdoor biergärtens. These vessels may be made of many materials, other than glass or stoneware, such as porcelain and metal. They may

also be seen in sizes larger than one liter, though vessels containing three or five liters of beer are impractical for drinking. Steins, especially those made of porcelain, are often elaborately decorated, and marketed as presentation pieces. These vessels, especially those made in the nineteenth century or earlier can be quite exquisite, very expensive, and are prized by collectors who rarely, if ever, actually use them.

Unique in the world of beer glassware is Belgium, where nearly every beer or brewery has its own distinctive glass. These range from the general shape of a willibecher, to that of chalices, goblets and snifters. Hoegaarden uses a heavy, hexagonally-faceted, straight-sided tumbler that is similar in profile

to an Anglo-American pint glass. Various abbey and Trappist beers have specific chalice-like glassware. Golden ales are served in stemmed tulip glasses of various shapes. Some lambic brewers specify porcelain or stoneware vessels, while others require that their beer be served in narrow flutes to allow an appreciation of the color.

One of Belgium's most unique glasses is that of Kwak, from Brouwerij Bosteels, which is served in a fluted tube glass with a spherical bulb at the base. Because this base is a sphere, it cannot stand up without being held in place with a wooden frame, which is provided, adding to the drama of the presentation. Originally, the beer was served to coachmen, who had a rack for their glass on the side of their coach seat.

The ten-inch Kwak glass is a short version of the classic English *yard of ale*, which was introduced in or before the eighteenth century. As the name implies, the yard was a three-foot tube with a sphere at the base that contained around 2.5 imperial pints of beer. As with the Kwak glass, it was reportedly designed for coachmen, though it has been more commonly used in recent years as part of drinking games.

Another vessel used for similar games is the German *boot*, glassware with a volume often larger than two liters, which is literally shaped like a boot. Skill in drinking a yard or a boot is demonstrated by the drinker who avoids a sudden face full of beer from the bulb of the yard or the toe of the boot when the glass is tipped above horizontal. With care, this skill can be mastered, but novices typically fail, with results which their companions will invariably find hilarious.

**Above: A variety of Belgian glassware: The La Chouffe tulip; Palm's tulip; Rodenbach Grand Cru in a goblet; and similar chalices from the Maredsous and Steenbrugge abbey breweries.** (Courtesy of the breweries)

**Below: A Chimay chalice.** (Chimay)

# Index